SpringerBriefs in Reproductive Biology

More information about this series at http://www.springer.com/series/11053

SpringerBriefs in Reproductive Biology is an exciting new series of concise publications of cutting-edge research and practical applications in Reproductive Biology. Reproductive Biology is the study of the reproductive system and sex organs. It is closely related to reproductive endocrinology and infertility. The series covers topics such as assisted reproductive technologies, fertility preservation, in vitro fertilization, reproductive hormones, and genetics, and features titles by the field's top researchers.

Ashok Agarwal • Damayanthi Durairajanayagam
Gurpriya Virk • Stefan S. Du Plessis

Strategies to Ameliorate Oxidative Stress During Assisted Reproduction

 Springer

Ashok Agarwal
Center for Reproductive Medicine
Cleveland Clinic
Cleveland, OH, USA

Gurpriya Virk
Melbourne, VIC, Australia

Damayanthi Durairajanayagam
Discipline of Physiology
    Faculty of Medicine
MARA University of Technology
Kuala Lumpur, Malaysia

Stefan S. Du Plessis
Division of Medical Physiology
Stellenbosch University
Stellenbosch, South Africa

ISSN 2194-4253   ISSN 2194-4261 (electronic)
ISBN 978-3-319-10258-0   ISBN 978-3-319-10259-7 (eBook)
DOI 10.1007/978-3-319-10259-7
Springer Cham Heidelberg New York Dordrecht London

Library of Congress Control Number: 2014948773

Printed on acid-free paper

Springer is part of Springer Science+Business Media (www.springer.com)

Preface

Assisted reproduction has become a very common treatment option for couples dealing with infertility. Despite the successes of these *in vitro* technologies, it is impossible to fully reproduce the *in vivo* environment. Not only are the gametes and embryos exposed to unnatural conditions, but they are also manually manipulated. All of these interventions can lead to the generation of excess reactive oxygen species. Under such conditions, natural antioxidant defences cannot preclude these pathological ROS levels. Ultimately, this leads to the development of oxidative stress which affects gamete survival, fertilization and embryogenesis negatively. It is therefore imperative to find and pursue solutions, such as antioxidant treatment, that may help ameliorate this process.

Antioxidant therapies in ART are written by leaders in the field of oxidative stress in both male and female reproductive medicine. This manuscript aims to bridge the gap between basic research, the role of the embryologist as well as the clinician with regard to antioxidant treatment both *in vivo* and in the ART laboratory. It eloquently covers all aspects of ROS generation (both endogenous and exogenous) as well as the complex interplay of oxidative stress in the ART setting. Various antioxidants and their plausible therapeutic effects to minimize ROS and oxidative stress are also comprehensively discussed.

Cleveland, OH, USA   Ashok Agarwal
Kuala Lumpur, Malaysia   Damayanthi Durairajanayagam
Melbourne, VIC, Australia   Gurpriya Virk
Stellenbosch, South Africa   Stefan S. Du Plessis

Contents

List of Figures

List of Tables

Authors

Ashok Agarwal is a Professor at the Lerner College of Medicine, Case Western Reserve University and the Head of the Andrology Center and Director of Research at the Center for Reproductive Medicine, Cleveland Clinic, USA. He has researched extensively on oxidative stress and its implications on human fertility and has published over 500 original research articles and reviews. He serves on the editorial boards of several key journals in human reproduction and has edited over 20 medical text books/manuals related to male infertility, ART, fertility preservation, DNA damage and antioxidants. Ashok's current research interests are the study of molecular markers of oxidative stress, DNA fragmentation and apoptosis using proteomics and bioinformatics tools, as well as fertility preservation in patients with cancer, and the efficacy of certain antioxidants in improving male fertility.

Stefan S. du Plessis is a Professor and the Head of Medical Physiology at Stellenbosch University, South Africa, where he also heads the Reproductive Research Group. His research interests include factors that influence male gamete function and he has published more than 70 scientific articles and book chapters. Dr. du Plessis serves as editorial board member for two leading journals, is an NRF-rated researcher, and is currently a Fulbright Researcher at the Cleveland Clinic's Center for Reproductive Medicine.

Damayanthi Durairajanayagam, Ph.D. is a Senior Lecturer in Physiology at the Faculty of Medicine, MARA University of Technology, Malaysia. She received a Fulbright Scholarship for her research at the Center for Reproductive Medicine, Cleveland Clinic. Her research interests include oxidative stress and the use of proteomics and bioinformatics in studying the molecular markers of oxidative stress in infertile males.

Gurpriya Virk has a Master's in Clinical Embryology from Monash University, Melbourne, Australia. She did her research training at the Center for Reproductive Medicine, Cleveland Clinic. She is interested in investigating the role of oxidative stress in assisted reproduction.

Chapter 1
Introduction

Natural conception involves a series of intricate and well-orchestrated physiological processes in both male and female partners to ultimately culminate in a successful pregnancy. While conceiving a baby may come naturally for the majority of couples, about 15 % of all couples face infertility and achieving a pregnancy is very likely to require clinical intervention. Infertility is defined as the inability to conceive or carry a pregnancy to full term, after a year of regular, unprotected sexual intercourse. Assisted reproductive technology (ART) has been gaining popularity over recent years as it offers hope and a means for couples with less than ideal reproductive parameters or situations to conceive a baby. In vitro fertilization (IVF) and intracytoplasmic injection (ICSI) are two of the commonly used interventions in ART. While these interventions attempt to mimic as closely as possible the environment and conditions under which natural fertilization takes place, the exact same conditions cannot be recreated precisely in the laboratory setting. In fact, there are multiple factors during the various steps in ART that can hinder its outcome. One of the key causes of defective gametes, or non- or poorly-developing embryos in ART is the development or presence of oxidative stress while handling the gametes or embryos during the procedure.

Attempting to conceive through assisted reproduction is not only emotionally and physically taxing, but also carries a large financial cost. Thus, the goal for the couple going into the chosen technique would be to improve the quality of their gametes as much as possible, and simultaneously minimize the effect of oxidative stress (OS) that may be incurred during the mechanical manipulation of the gametes and embryos during the procedure, in order to optimize the ART outcome. This could be done by performing deliberate measures to reduce any incidental development of OS as well as to boost the antioxidant capacity of the gamete to withstand the detrimental assault of oxidation. Some of the exogenous sources/factors that can contribute to the generation of OS include the use of cryopreserved spermatozoa, exposure of the gamete to visible light, centrifugation, high temperatures, culture media and handling time, gamete manipulation, and incubator environment.

© The Author 2014
A. Agarwal et al., *Strategies to Ameliorate Oxidative Stress During Assisted Reproduction*, SpringerBriefs in Reproductive Biology,
DOI 10.1007/978-3-319-10259-7_1

Multiple research studies have previously investigated and reported the effects of various enzymatic and non-enzymatic antioxidant therapies on sperm parameters, fertilization rate, blastocyst development, embryo transfer and pregnancy outcome. Thus, the aim of this monograph is to review the current antioxidant strategies and how it can be used in a clinical setting to minimize the damages to the gamete and/or embryo caused by OS during ART.

Chapter 2
Sources of ROS in ART

Reactive oxygen species (ROS) are molecules derived from oxygen, which occur as a by-product of cellular oxidative respiration. ROS are free radical molecules containing one or more unpaired electrons in their outer shell. They tend to remove an electron from surrounding molecules to complete their octet. As a ROS stabilizes itself with the addition of an electron, the molecule from which it removed the electron now becomes a free radical. In this manner, a self-propagating chain reaction that produces ROS is generated. As ROS goes on to react with other molecules, the molecule undergoes structural and functional modifications (Sharma and Agarwal 1996).

Commonly-occurring ROS in basal conditions, such as superoxide anion (O_2^-), hydrogen peroxide (H_2O_2) and hydroxyl radical (OH^-), are powerful oxidants that are found in low concentrations in the genital tract of both males and females (Guerin et al. 2001; Agarwal and Allamaneni 2004). In the electron transport chain, oxygen functions as the final electron acceptor. Most oxygen ions bind with hydrogen ions to form water molecules, but oxygen ions that do not undergo this reaction, are released from the mitochondria in the form of free radicals. As levels of ROS increase, the cell's antioxidant defences become overwhelmed and are inadequate in scavenging the unstable metabolites of oxygen, leading to a state of oxidative stress (OS) (Pasqualotto et al. 2004).

A subset of ROS is the nitrogen-containing compounds, known as reactive nitrogen species (RNS), whose formation is catalyzed by nitric oxide synthase enzymes (O'Bryan et al. 1998). Examples of RNS include nitric oxide ˙NO, nitroxyl ion HNO, peroxynitrate anion $ONOO^-$ and nitrosyl-containing compounds. Physiological levels of RNS helps maintain normal sperm parameters and general reproductive functions, as well as stimulate the immune functions. However, pathological levels of RNS contribute towards a state of nitrosative stress, which negatively impacts sperm function and its fertilizing capacity, resulting in compromised male reproductive function (Doshi et al. 2012).

In the ART setting, the effect of OS is augmented by the lack of endogenous physiological defence mechanisms and the multiple potential sources of ROS, both

© The Author 2014
A. Agarwal et al., *Strategies to Ameliorate Oxidative Stress During
Assisted Reproduction*, SpringerBriefs in Reproductive Biology,
DOI 10.1007/978-3-319-10259-7_2

internally [endogenously from the gametes and exogenously from the environment and by manipulation of the gamete/embryo] and externally [*in vitro* factors such as oxygen partial pressure and culture media], during the ART procedure. As the *in vitro* setting during ART is unable to completely mimic physiological conditions *in vivo*, the gametes used in assisted reproduction are more susceptible to the detrimental effects of oxidative stress. This negatively impacts on the ART outcome (Lampiao 2012). Among the external factors that could potentially affect the gamete/embryo viability *in vitro* are cryopreservation/freeze-thaw procedure, visible light, the specific ART technique employed, pH and temperature, centrifugation, culture media, and most importantly, oxygen concentration or partial pressure. These exogenous ROS-inducing factors, along with endogenous sources of ROS are depicted in Fig. 2.1 and will be discussed in the following sections.

2.1 In Vivo Generation of ROS

2.1.1 *In Vivo Generation of ROS in the Male*

Spermatozoa use ATP as a source of energy, which is produced through mitochondrial oxidative phosphorylation and glycolysis. Spermatozoa physiologically undergo both aerobic and anaerobic metabolic processes, both of which contribute towards the production of ROS (du Plessis et al. 2008). Under physiological conditions, the primary source of ROS is the escape of activated oxygen from the mitochondria during oxidative respiration. Under typical conditions *in vivo*, ROS is neutralized continuously to maintain the required levels of ROS to facilitate the normal functioning of human spermatozoa. The low levels of ROS aid in the physiological processes that maintains male reproductive functions such as cell signalling, regulation of tight junctions, steroidogenesis, capacitation and acrosome reaction, sperm motility, and zona pellucida binding (de Lamirande and Gagnon 1993; Doshi et al. 2012).

Besides the ROS generated intrinsically from the plasma membrane and mitochondria of spermatozoa (Gavella and Lipovac 1992), other cells may also produce ROS in the male genital tract. Human spermatozoa generate O_2^-, which spontaneously dismutates to H_2O_2 (Alvarez et al. 1987).

Impaired spermatogenesis leads to the production of immature and morphologically-abnormal spermatozoa, which contribute to ROS in the ejaculate. A major extrinsic source of ROS in the human ejaculate is leukocytes. These may be present in the ejaculate due to *in vivo* inflammatory processes. Leukocytes physiologically produce about a 100-fold more ROS than spermatozoa (Plante et al. 1994; de Lamirande and Gagnon 1995). High levels of ROS produced during leukocytospermia, play a major role in infection, inflammation and cellular defence mechanisms against pathogens.

The highly reactive ROS combines easily with molecules causing cellular damage (de Lamirande and Gagnon 1995; Agarwal et al. 2005b). The plasma membrane of the spermatozoon contains a high amount of polyunsaturated fatty

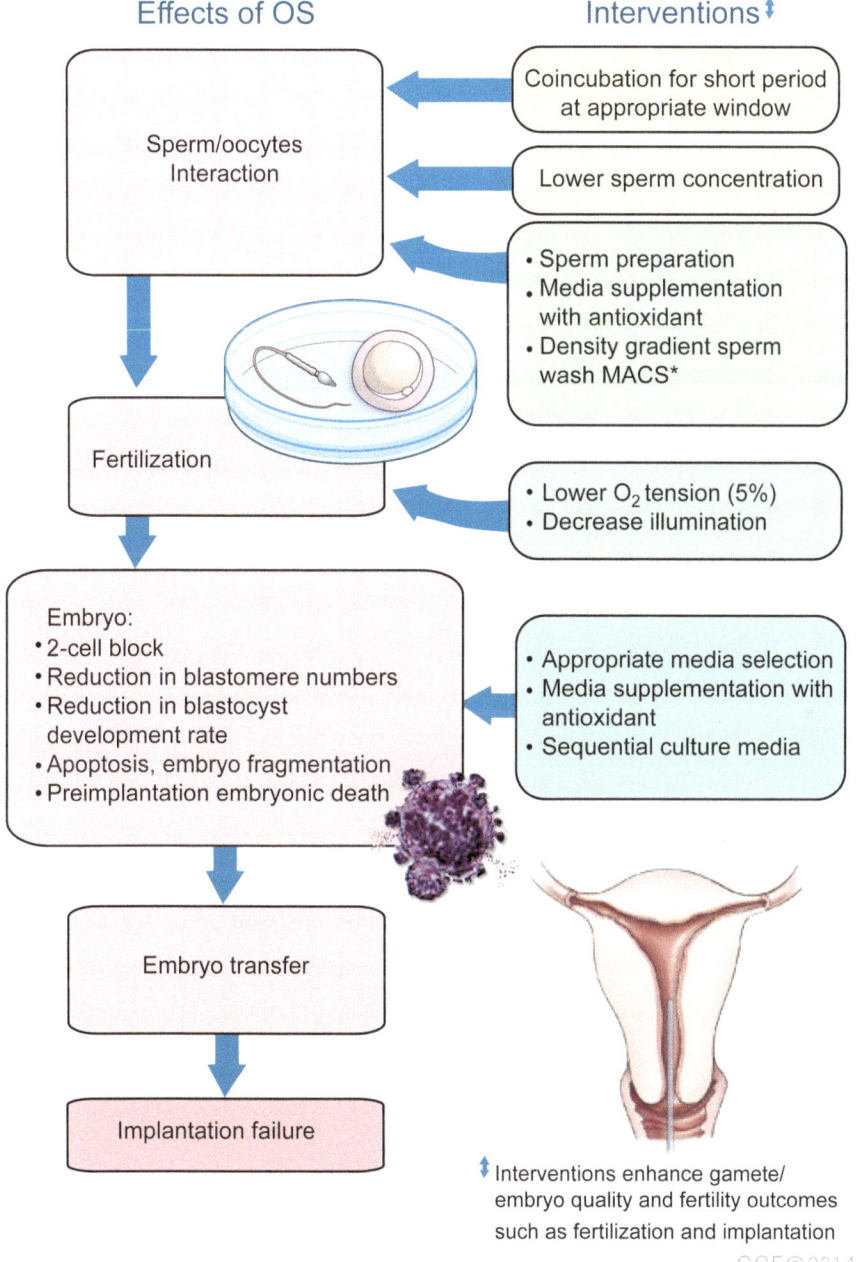

Fig. 2.1 Sources of reactive oxygen species in assisted reproductive technology
MACS magnetic cell separation

acids (PUFAs), making it very susceptible to elevated ROS levels during OS. The double bonds of membrane lipids are easily oxidized by ROS, causing a decrease in membrane fluidity. Besides the lipid membranes, OS also affects the structural proteins and nucleic acids of the sperm. As a result of this lipid peroxidation, the motility of the spermatozoon is compromised and could ultimately lead to immobilization of spermatozoa (Agarwal et al. 2003). OS leads to DNA fragmentation and damage in the nucleus of spermatozoa. Typically this include single- or double-strand breaks, DNA crosslinks, chromosomal rearrangements as well as deletions (Aitken and Krausz 2001). Damage to the sperm DNA by the increased levels of ROS is implicated in adverse outcomes such as higher incidence of abortion and childhood cancers (Baker and Aitken 2005). High ROS levels also cause a decrease in the mitochondrial membrane potential, which is an initiating event in the apoptosis cascade (Wang et al. 2003).

The consequences of high ROS levels are especially significant in infertile patients seeking assisted reproduction as sperm selected from these patients are very likely to have already been exposed to OS (Saleh et al. 2003). Increased OS in spermatozoa have a negative impact on ART outcome, with poor fertilization rates, embryo development and pregnancy rates (Baker and Aitken 2005).

The pro-oxidant activities of ROS are quenched by circulating antioxidants in order to prevent the detrimental effects of OS on the male gamete. Antioxidants function to maintain low, physiological concentrations of ROS in the cell and act as a natural defence mechanism against OS. Antioxidants can either be enzymatic (e.g. superoxide dismutase SOD, catalase, glutathione peroxidase GPx and glutathione reductase GR) or non-enzymatic (e.g., vitamins C, E and A, glutathione, pyruvate, taurine and hypotaurine, albumin and transferrin) (Sharma et al. 2004; Sikka 2004).

The mid-piece of the human sperm contains mainly enzymatic antioxidants (SOD, GPx and GR) and the plasma membrane contains a few non-enzymatic antioxidants (vitamins E and A, transferrin); while the seminal plasma contains both enzymatic and non-enzymatic antioxidants (Agarwal and Prabakaran 2005). These antioxidants contribute towards the total antioxidant capacity of the gamete. Under normal conditions, seminal plasma has adequate antioxidant mechanisms to minimize ROS action (Ford et al. 1997). However, during preparation of the sperm for the assisted reproductive technique, the sperm is separated from the seminal plasma, which decreases the spermatozoa's natural defence mechanisms.

With time, the limited antioxidant capacity of the spermatozoa declines as the aging sperm has reduced GPx and SOD activity (Sikka 2004). While ROS is present in sperm from both fertile and infertile men, more infertile men tend to have excessive levels of ROS levels compared to fertile men. Elevated ROS production compromises the structural integrity and functional capacity of the sperm, causing lipid peroxidation, DNA fragmentation and apoptosis. Elevated RNS levels can also cause lipid peroxidation, DNA damage, inhibition of steroidogenesis, increased caspase activity ultimately leading to apoptosis. Furthermore, RNS negatively impacts sperm parameters by reducing motility and viability, causing abnormal morphology, decreasing capacitation and the acrosome reaction, as well as sperm-oocyte fusion (Doshi et al. 2012).

In summary, spermatozoa have increased susceptibility to OS, as there is a lack of cytoplasm in the mature sperm compared to somatic cells, and therefore its poorer antioxidant capacity tends to render the spermatozoa at greater vulnerability to OS.

2.1.2 In Vivo Generation of ROS in Females

The role of ROS in female infertility has been of significant interest and research over the last decade. The presence of ROS in the fluids and organs involved in female reproductive processes has been well documented. In the female, the sources that generate ROS *in vivo* are mainly the follicular fluid, fallopian tube and uterine environment (Pasqualotto et al. 2004; Bedaiwy et al. 2004; Guerin et al. 2001). At these locations, ROS appear to have a physiological role in oocyte maturation, ovarian steroidogenesis and ovulation, implantation and formation of fluid-filled cavity, blastocyst, luteolysis and luteal maintenance in pregnancy. Moreover, a limited amount of lipid peroxidation relevant to ROS in follicular fluid was found obligatory for establishing pregnancy in human IVF cycles (Agarwal et al. 2005a).

Each month, a cohort of oocytes develops and matures in the ovary. However, meiosis I will resume only in the dominant oocyte. This process could be threatened by an increase in ROS. Conversely, high levels of ROS could be inhibited by antioxidants. This implies an intricate association between ROS and antioxidants in the ovary (Behrman et al. 2001; Agarwal et al. 2012). Inside the ovary, the follicular fluid environment enclosing the oocyte plays an important role in the fertilization process. Once fertilized, follicular fluid (along with the fallopian tube and endometrium), plays an important role in subsequent embryo development. During follicular maturation, oocytes are well-sheltered against lethal injury due to OS by antioxidants such as catalase, superoxide dismutase, glutathione transferase GST, paraoxonase PON, heat shock protein 27 and protein isomerase (Ambekar et al. 2013). Additionally, there are other antioxidants present such as vitamin E, carotene, ascorbate, cysteamine, taurine, hypotaurine, transferrin, thioredoxin, and dithiothreitol. The imbalance between these antioxidants and pro-oxidants in female infertility has been postulated to alter gene expression and impair adenosine triphosphate (ATP) generation (Agarwal et al. 2005a), the latter of which can affect ovulation, oocyte quality, fertilisation, embryo development and implantation (Agarwal et al. 2006a; Revelli et al. 2009; Das et al. 2006; Pasqualotto et al. 2009). An excess of free radicals also play a key role in gynaecological problems such as polycystic ovarian syndrome (PCOS), endometriosis and tubal factor infertility (Pasqualotto et al. 2009). Numerous studies discussed in this section have demonstrated the relationship between these disorders, OS and successful ART outcome.

The oocyte, granulosa and surrounding cells, such as the endothelial and thecal cells constitute the follicular fluid environment. It also contains phagocytic macrophages, parenchymal steroidogenic and endothelial cells that generate ovarian ROS (Halliwell and Gutteridge 1988; Agarwal et al. 2005a). The composition of follicular fluid is a key feature for predicting a successful ART outcome in females (Agarwal et al. 2005a). While large amounts of ROS in the follicular fluid pose a serious threat to the success of assisted reproduction, a limited amount of ROS was found to be obligatory for establishing pregnancy in human IVF cycles (Oral et al. 2006). A summary of the experimental studies investigating the relationship between follicular fluid and ART outcome is portrayed in Table 2.1.

In assisted reproduction, the assessment of ROS and RNS in the follicular fluid of women undergoing IVF is usually obtained by aspiration of follicular fluid from each follicle during the oocyte retrieval process. Follicular fluid samples are then centrifuged and evaluated, most commonly by the chemiluminescence assay using luminol (other methods include nitroblue tetrazolium staining, thiobarbituric acid-reacting substances (TBARS), and flow cytometry) (Askoxylaki et al. 2013). In a study by (Jana et al. 2010), the upper cut-off limit for ROS levels, beyond which viable embryo formation is not favourable, was observed to be approximately 107 counted photons per second (cps)/40 μL of follicular fluid. This cut-off level (previously determined in infertile women with tubal factor), was validated next in women with PCOS and endometriosis, in which they were shown to adversely affect oocyte and embryo development as well as pregnancy outcome. ROS values above this threshold value were shown to hinder oocyte quality, maturation, fertilization, and embryo formation. Conversely, significantly lowered levels of ROS (<100 cps) were linked with good embryo quality. Similar were also reported the lower and upper ROS values as 41 and 150 cps respectively (Das et al. 2006).

A prospective study by Attaran et al. (2000) reported that ROS levels were significantly lower in the follicular fluid of patients (with tubal disease, endometriosis and idiopathic infertility) who failed to establish pregnancy as compared to those who did (although the specific cut off values were not described). These results were in accordance with Pasqualotto's study, in which patients who became pregnant had higher lipid peroxidation levels and total antioxidant capacity (TAC) (Pasqualotto et al. 2004). Despite the fact that an imbalance of pro-oxidants and antioxidants can cause a disturbance in natural female reproductive tendencies, however these results indicate that physiological levels of ROS within the follicular fluid is essential for different phases of oocyte development and maturation. However, the exact function of ROS in the follicular fluid remains unknown (Pasqualotto et al. 2004; Oyawoye et al. 2003; Jana et al. 2010).

In a study of 63 women undergoing IVF, a total of 303 follicular aspirates were analysed using ferric reducing antioxidant power (FRAP) assay for baseline TAC (Oyawoye et al. 2003). The study revealed that TAC levels were significantly higher in follicular fluid where the oocyte was successfully fertilized. In a more recent study, the association between follicular fluid, ROS levels, TAC, ROS-TAC score and pregnancy following ICSI were evaluated in 138 women (Bedaiwy et al. 2012). Results of this study illustrated that pregnancy cycles were associated with significantly lower ROS and higher TAC. Interestingly, their study found that TAC levels were higher in women with endometriosis. Their ROS-TAC scores were also higher and were associated with a greater probability of having normal oocytes. In these women, elevated TAC levels, higher ROS-TAC scores and lower ROS levels in their follicular fluid were allied with a successful pregnancy after ICSI (Bedaiwy et al. 2012).

Table 2.1 Clinical parameters of studies examining reactive oxygen species in follicular fluid in infertile women

Study	Sample collected	Patient population/characteristics	Factors measured in the sample collected	Outcomes measured
Attaran et al. (2000)	FF	Patients undergoing ovarian stimulation for ART: women with tubal disease, male factor, endometriosis, idiopathic infertility, ovulatory dysfunction, pelvic adhesions	OS markers: 1. ROS 2. TAC	Age, number of oocytes recovered, percentage of oocytes fertilized, achievement of pregnancy
Barrionuevo et al. (2000)	FF	Patients undergoing IVF	1. Nitric oxide metabolites nitrate/nitrite (NO_3/NO_2) 2. IL-1β levels	Oocyte maturation, fertilization rate, embryo cleavage rate
Oyawoye et al. (2003)	FF	Patients undergoing IVF-ET treatment	1. TAC	Oocyte retrieval, fertilization rate, embryo viability
Pasqualotto et al. (2004)	FF	Patients undergoing IVF: male factor infertility, tubo-peritoneal factors, idiopathic infertility, ovulatory factors	OS markers: 1. LPO 2. TAC	Oocyte maturity, fertilization rate, cleavage rate embryo quality, and pregnancy rate
Das et al. (2006)	FF	Patients undergoing IVF-ET treatment by controlled ovarian stimulation (long protocol): tubal factor infertility	OS markers: 1. ROS 2. LPO	Oocyte quality and fertilization potential, embryo quality
Chattopadhayay et al. (2010)	FF	PCOS	OS markers: 1. ROS 2. LPO 3. TAC	Meiotic spindle formation, fertilization rate, number of good quality embryos, clinical pregnancy rates
Jana et al. (2010)	FF	Patients undergoing IVF-ET by controlled ovarian stimulation: tubal factor infertility, endometriosis, PCOS	OS markers: 1. ROS 2. LPO 3. DNA fragmentation 4. TAC	Oocyte quality, fertilization rate, embryo quality

(continued)

Table 2.1 (continued)

Study	Sample collected	Patient population/characteristics	Factors measured in the sample collected	Outcomes measured
Borowiecka et al. (2012)	FF	Patients undergoing IVF	Lipid and protein peroxidation markers: 1. TBARS 2. Protein carbonyl 3. Thiol groups	Pregnancy rates
Bedaiwy et al. (2012)	FF	Patients who had ICSI: couples with male infertility, endometriosis, tubal disease, idiopathic infertility, ovulatory dysfunction, combined male and female infertility	OS markers: 1. ROS 2. TAC	ROS-TAC score, number of follicles, number of oocytes retrieved, oocyte quality, pregnancy rate
Otsuki et al. (2012)	FF	Infertile patients	1. Redox state 2. Albumin	Oocyte viability
Rajani et al. (2012)	FF	Patients undergoing ICSI-ET: with endometriosis, PCOS, tubal infertility (control)	1. ROS levels	Presence of meiotic spindle, number of oocytes retrieved, mature MII oocytes, fertilization rate, good embryo formation rate, pregnancy rate
Singh et al. (2013)	FF	Patients undergoing IVF: endometriosis and tubal factor (controls)	OS markers: 1. ROS 2. NO (nitrite/nitrate) 3. LPO 4. TAC 5. Antioxidant (enzymatic and non-enzymatic) levels	Oocyte quality, embryo quality, pregnancy rate

Reference	Sample type	Patient group	OS markers	Outcome measures
Liu et al. (2013)	FF	Patients undergoing IVF-ET: endometriosis, tubal factor infertility	OS markers: 1. ROS 2. SOD 3. Vitamin E	Oocyte quality, fertilization rate
Seino et al. (2002)	GC	Patients undergoing IVF-ET treatment and ICSI: endometriosis, male factor, tubal factor, unknown	1. 8-OHdg expression in granulosa cells	Oocyte quality, fertilization rate, embryo quality (good embryo rate)
Liu and Li (2010)	GC	Patients undergoing IVF-ET: tubal factor infertility	1. MDA 2. SOD 3. Apoptosis 4. Good embryo rate	Number of retrieved oocytes, oocyte maturity, embryo quality, fertilization, cleavage
Karuputhula et al. (2013)	GC	Patients undergoing IVF-ET: endometriosis, PCOS, tubal factor infertility	OS markers: 1. ROS 2. MMP 3. DNA fragmentation 4. Apoptosis	GC characteristics: Fertilization rate, number of oocytes retrieved, oocyte quality, good quality embryo formation rate, pregnancy outcome
Polak et al. (2001b)	PF, plasma samples	Infertile women with minimal or mild endometriosis, unexplained infertility, PCOS, tubal infertility	1. 4-HNE levels 2. MDA levels (lipid peroxide levels)	Oxidative stress/free radicals activity
Bedaiwy et al. (2004)	Culture media	Patients undergoing IVF/ICSI: male factor, anovulation, endometriosis, tubal factor, unexplained and other factors	1. ROS in culture media	Fertilization rate, cleavage rate, fragmentation, embryonic fragmentation levels, blastocyst formation rate

Sample types: *4-HNE* 4-hydroxynonenal, *8-OHdG* 8-hydroxy-2′-deoxyguanosine, *ART* assisted reproductive technology, *CoQ$_{10}$* Coenzyme Q$_{10}$, *DNA* deoxyribonucleic acid, *EDTA* ethylenediaminetetraacetic acid, *FF* follicular fluid, *GC* granulosa cells, *hCG* human chorionic gonadotrophin, *ICSI* intracytoplasmic sperm injection, *ICSI-ET* intracytoplasmic sperm injection and embryo transfer, *IL-1β* Interleukin-1 beta, *IVF* in vitro fertilization, *IVF-ET* in vitro fertilization and embryo transfer, *LPO* lipid peroxidation, *MDA* malondialdehyde, *MII* metaphase II, *MMP* mitochondrial membrane potential, *NO* nitric oxide, *OS* oxidative stress, *PCOS* polycystic ovarian syndrome, *PF* peritoneal fluid, *ROS* reactive oxygen species, *SOD* superoxide dismutase, *TAC* total antioxidant capacity, *TBARS* thiobarbituric acid reactive substances

In a much earlier study, Polak's group demonstrated the total antioxidant status of peritoneal fluid in infertile women (Polak et al. 2001a). The peritoneal fluid was obtained from infertile women distressing from minimal or mild endometriosis, unexplained infertility, tubal infertility and a few fertile women. The results of this study demonstrated a significantly lower antioxidant status in peritoneal fluid obtained from women with unexplained infertility.

Oocyte quality is another important determinant of IVF outcome. The follicular fluid aspirated during oocyte retrieval from women with endometriosis and tubal infertility undergoing IVF was measured using spectroscopy and HPLC (Singh et al. 2013). Increased levels of ROS and NO in endometriosis and tubal infertility were found to correlate with poor oocyte and embryo quality. Further, increased levels of ROS, nitric oxide, lipid peroxidation, cadmium and lead were seen in women who failed to become pregnant compared to those who did.

Borowiecka et al. (2012) evaluated the levels of lipid and protein peroxidation markers (TBARS, protein carbonyl, and thiol groups) in the follicular fluid of patients undergoing IVF. The OS markers were compared between the pregnancy positive and pregnancy negative patient groups. Results demonstrate that the mean follicular fluid TBARS level of non-pregnant women was significantly higher than that found in pregnant women. These findings suggested that elevated follicular fluid lipid and protein peroxidation levels may have a negative impact on IVF outcome and also supported the idea that increased levels of OS markers in follicular fluid may play an important role in fertility.

The association between malondialdehyde (MDA), superoxide dismutase (SOD) and incidence of apoptosis of granulosa cells in follicular fluid was examined in women with tubal factor infertility (Liu and Li 2010). The level of MDA and the activity of the SOD were measured by TBARS using a chemiluminescence methods, respectively. It was discovered that non-pregnant patients showed significantly higher MDA levels, higher incidence of apoptosis and lower SOD levels in the granulosa cells with lower good-embryo rate as compared to the pregnant patients. In general, OS induced apoptosis in granulosa cells and subsequently lowered oocyte quality, leading to poor outcome of IVF-ET. The levels of 8-hydroxy-2′-deoxyguanosine (8-OHdG) in granulosa cells were detected in Seino's study (Seino et al. 2002). The outcome of their study suggested that OS in granulosa cells reduced fertilization rates and consequently reduced embryo quality. They also found that the quality of oocytes of endometriosis patients was reduced by the presence of 8-OHdG. The outcome of this study was in agreement with that of Karuputhula's group. A ~20- and 100-fold increase in granulosa cells ROS generation and MMP, was seen in PCOS patients as compared to tubal factor patients. Significant apoptosis was also evident in PCOS and endometriosis patients. IVF outcome parameters comprise fertilization rate, formation rate of good quality embryo, and pregnancy rates, and these were badly affected in endometriosis patients (Karuputhula et al. 2013).

Another study which was performed on patients with PCOS, examined the meiotic spindle in oocytes along with ROS levels in follicular fluid of women (Rajani et al. 2012). It was demonstrated that women with endometriosis had low ROS levels and good spindle imaging results suggesting a possible role of endometrial receptivity accounting for lower pregnancy rates in these women. Poor oocyte quality, as reflected by higher mean ROS levels and low number of oocytes with spindle visualization, could be the factor impeding pregnancy in women with PCOS as compared to women with tubal block (Rajani et al. 2012). Another group examined that the effect of follicular fluid OS on the formation of meiotic spindle in oocytes and the outcome in women with PCOS (Chattopadhayay et al. 2010). Oocytes were examined to visualize for the meiotic spindle using a PolScope (a polarised light microscope). From the results of this study, it was observed that OS was responsible for the absence of the meiotic spindle which was significantly found in the groups with low fertilization rate, reduced number of good quality embryos and clinical pregnancy. Otsuki's group studied the influence of the redox state of follicular fluid on the viability of aspirated human oocytes (Otsuki et al. 2012). The redox state of the follicular fluid and serum, at the time of oocyte retrieval, was analysed by high performance liquid chromatography. Their results showed that the redox state of follicular fluid that contained degenerated oocytes had a significantly higher oxidised state compared with fluids that capitulated normal oocytes. A prospective case-control study was conducted in endometriosis patients who underwent IVF-embryo transfer IVF-ET. The expression and the role of OS markers in serum and follicular fluid of the patients were investigated. The results of the study reported significantly higher levels of ROS in both serum and follicular fluid. Mature oocyte and fertilization rates were also significantly lower than in the control group (Liu et al. 2013).

Nitric oxide concentrations in follicular fluid have been negatively associated with low pregnancy rates. It was found that serum nitric oxide concentrations in patients with tubal factor or peritoneal factor infertility negatively correlated. They reported that higher concentrations of nitric oxide are associated with implantation failure, which then result in lower pregnancy rates (Bedaiwy et al. 2004).

In summary, the findings of the studies discussed in this section highlight the importance for the detection of ROS in the female reproductive system for a variety of reasons and indicate how ROS levels can also be used to determine the link between gynaecological diseases (PCOS, endometriosis, tubal factor) and OS.

2.2   In Vitro Generation of ROS in ART

2.2.1   Cryopreservation

Cryopreservation is a process whereby cells and whole tissues are conserved by cooling to sub-zero temperatures (-196 °C) (Di Santo et al. 2012). Recent advances in assisted reproduction and embryology have made cryopreservation an appropriate method for long-term storage of human reproductive cells, embryos and gonadal tissues to preserve and protect fecundity in cases of infertility, malignancy (Woods et al. 2004) and in some non-malignant treatments (such as diabetes and autoimmune disorders that may lead to testicular injury) (Anger et al. 2003). Nonetheless, despite highly optimised protocols appear to improve cell viability, the extreme stress of freezing and thawing treatments can modify the structure and integrity of the sperm plasma membrane (mainly composed of phospholipids and cholesterol) (Giraud et al. 2000; Di Santo et al. 2012).

Various reports studied the relationship between cryopreservation and antioxidant defence system. The occurrence of DNA fragmentation in testicular sperm of men with obstructive azoospermia, such as in vasectomised males, those with blocked or missing ducts, and those with an absence of the vas deferens was investigated (Dalzell et al. 2004) The study showed a significant increase in DNA fragmentation at 24 h after incubation of fresh testicular sperm from men with obstructive azoospermia. Frozen-thawed sperm DNA was found to be significantly more damaged than fresh testicular sperm DNA. Even after 4 and 24 h post-thaw incubation, sperm DNA continued to become more damaged compared to fresh sperm DNA. These findings were further confirmed by Thomson's study, which reported increased levels of 8-oxo-7,8-dihydro-2′-deoxyguanosine (8-OHdG) (a biomarker of oxidative stress), indicating an increase in sperm DNA fragmentation due to cryopreservation of human semen (Thomson et al. 2009). Furthermore the effect of cryopreservation on motility and viability was evaluated on spermatozoa of men struggling with infertility. The results in these men showed a significant decrease in sperm motility and viability post-cryostorage. In addition, cryopreservation/thaw significantly increased sperm DNA fragmentation and DNA oxidative damage (Zribi et al. 2010).

In human oocytes, cryopreservation frequently leads to developmental arrest during early cleavage stages and display aberrant patterns of cytokinesis (cytoplasmic division of a cell at the end of mitosis or meiosis, bringing about the separation into two daughter cells) (Van Blerkom and Davis 1994). Various studies (Gualtieri et al. 2009; Jones et al. 2004) revealed significant reduction of mitochondrial potential in slow cooled human oocytes resulted from changes in redox status of the cell due to cryopreservation. However, a study by Chen et al. (2012) evaluated the impact of vitrification on mitochondrial membrane potential (MMP) in human metaphase II (MII) oocytes, and the changes of mitochondrial membrane potential in thawed MII oocytes. It was observed that in the vitrification/thawing process, the mitochondrial membrane potential of MII oocytes could have temporary dynamic changes within a couple of hours post-thaw but would recover fully after 4 h of culture. In another study, the effect of different vitrification protocols on ROS were evaluated in human ovarian tissues, which were exposed to different vitrification solutions (Rahimi et al. 2003).The intracellular redox state levels were measured using the fluorescent dye dichlorodihydrofluorescein diacetate. After vitrification and warming, apoptotic cells imaging was monitored by anti-caspase-3 immunolabelling. The results showed that a slower cooling of tissue resulted in significantly elevated ROS levels and apoptosis after warming (Rahimi et al. 2003).

Based on the outcome of multiple studies, there appears to be a clear picture where cryopreservation is responsible for OS during ART. The higher incidence of studies regarding the male gamete may likely confirm that spermatozoa are more vulnerable to redox alterations than oocytes during cryopreservation. This difference in sensitivity to OS is due to the higher susceptibility of spermatozoa to lipid peroxidation and to the limited amount of ROS scavengers available, as the antioxidant defences are localised mainly in the seminal plasma (Bozhedomov et al. 2009).

2.2.2   Visible Light

Visible light, also known as visible spectrum, is the portion of the electromagnetic spectrum that is visible to the human eye, having a wavelength in the range of 400–700 nanometres (nm)—between the infrared light with longer wavelengths and the ultraviolet light with shorter wavelengths (Buser et al. 1992) (Fig. 2.1). In an ART laboratory, oocytes, zygotes and embryos which are kept in an artificial medium for assisted fertilization procedures such as IVF and ICSI, are exposed to daylight or artificial light for variable periods before the embryo is transferred to females.

Over the years, enormous consideration has been given to the role of visible light in the production of ROS in animals (Takenaka et al. 2007; Moshkdanian et al. 2011). However, the effects of light have not been studied as extensively in humans. Light is either measured as units of intensity (lux) or by the level of irridation (W/m^2). According to an older study, rabbit oocytes when exposed to strong "cool white" fluorescent light of 3,250 lux for 20–30 min at 37 °C developed into normal near-term fetuses. However, this observation does not mean that visible light is undisruptive to the gametes and embryos of humans (Bedford and Dobrenis 1989).

In an IVF laboratory, light is generated by microscopes, fluorescent lighting and indirect sunlight. When gametes and embryos are exposed to light, it is absorbed by intracellular chromophores, which includes plasma membrane NADPH oxidase system consisting of flavoproteins and cytochrome b (Edwards and Silva 2001; Eichler et al. 2005). These chromophores are photosensitizers—which means they are able to absorb light and then transfer the energy to nearby oxygen molecules. Electrons in the shell get excited and shift to a shorter-lived singlet state. These excited electrons then change their spin and produce a longer-lived triplet state. As it returns to the ground state, it releases energy. This energy is then transferred to oxygen to generate ROS (singlet oxygen and free radicals) that mediate cellular toxicity (Girotti 2001). This mechanism of cellular toxicity was also described earlier (Hockberger et al. 1999). It was shown that violet-blue light (445–455 nm), irradiated from an inverted fluorescence microscope, initiated photoreduction of flavins. These activate flavin-containing oxidases in mitochondria and peroxisomes, resulting in production of H_2O_2 in human foreskin keratinocytes (HK cells).

The intensity and spectral composition of light reaching embryos during IVF was also investigated, and it was detected that microscopes, at a setting appropriate for embryo inspection, produced light at 2,500–5,000 lux. Meanwhile, light from other sources like natural room lighting and sunlight were more than tenfold lower, at 200–400 lux and had little effect on cultured embryos (Ottosen et al. 2007). It was suggested that microscope light exposure is a more hazardous source of light for oocyte and embryo viability compared to ambient light. The light intensity and wavelengths during embryo manipulation are important factors to maintain pre-implantation embryos viability *in vitro*.

The effect of visible light on human sperm motility and hyperactivation was demonstrated in a recent study (Shahar et al. 2011). The group evaluated the pathways mediating these effects. In their experiment, they irradiated human spermatozoa for 3 min with 40 mW/cm^2 visible light (400–800 nm with maximum energy at

600 nm) and suggested that ROS was produced during incubation, and that the production was enhanced after 1–3 min of light irradiation.

Over the years, attention has been focused towards the light microscopes used in IVF laboratories. Scientists recommend reducing the inspection light to as low as possible and maintaining the exposure time to as short as possible. In addition, microscope filters (optical filters) mounted on the microscope are used in a variety of microscopy applications for increasing contrast, obstructing ambient light, and eliminating harmful ultraviolet or infrared light (green filters block wavelengths below 500 nm). Moreover, to reduce ultraviolet (UV) lightwaves which are emitted from fluorescent lights, the fluorescent light could be substituted with yellow lights or by positioning UV light protectors over the fluorescent tubes (Ottosen et al. 2007).

2.2.3   ROS from Gamete Manipulation/ART Technique

In an *in vitro* environment, handling or manipulating the gametes and embryos during ART brings forth a risk of exposing these cells to higher than physiological levels of levels ROS (Lampiao 2012).

During ART procedures, the gametes are manipulated and prepared for various fertilization procedures such as IVF and ICSI. During conventional IVF, the selected samples of spermatozoa (concentration in millions) and oocytes are combined in the fertilisation medium in a petri dish and inspected for fertilisation after 16–20 h of incubation. During this period, ROS are generated from sources like oocytes, the cumulus cell mass and spermatozoa in culture medium. On the other hand, during the ICSI procedure, the cumulus cells are stripped from around the oocyte and a single sperm is injected directly into an oocyte before incubation in culture medium. Thus in an ICSI procedure, the oocyte and spermatozoa are the only potential sources of ROS. Furthermore, this procedure involves a shorter incubation time, which decreases the exposure of gametes to external environmental factors (Agarwal et al. 2006b).

However, since most ICSI cycles are done due to poor sperm characterises (which is not the case in an IVF procedure), therefore, it is commonplace to find higher rates of spermatozoa with greater DNA fragmentation levels in ICSI compared to IVF cycles (Benchaib et al. 2003). In addition, it was hypothesised that during the IVF procedure, human zona pellucida has the capacity to select against sperm with aneuploidy, which was supported by a theory that IVF steps leads to a 'natural selection of spermatozoa' (Van Dyk et al. 2000). The spermatozoon that is chosen for fertilization will have normal morphology and be exceedingly motile with intact DNA (Benchaib et al. 2003). Also with ICSI, in case of very poor sperm characteristics, the choice of spermatozoa to be injected is done using a very imperfect criteria and where it is tough to choose one normal motile sperm, so the risk of injecting spermatozoa with ROS and impaired DNA is high (Gandini et al. 2000; Irvine et al. 2000). It was also assumed that during the ICSI procedure, the selection of a motile and morphologically normal spermatozoa is operator-dependant. Therefore this spermatozoon has additional chances of having intact DNA, but at the same time a spermatozoa can be considered as 'normal' and still have damaged DNA (Host et al. 2000).

Elevated ROS levels were observed in the follicular fluid of women, whereby high ROS levels were prominent in those undergoing ICSI cycles, but not in IVF (Lee et al. 2012). It was demonstrated that the ICSI procedure might induce stress or shearing force on the plasma membrane of the oocyte, while the integrity of the plasma membrane has been considered as the origin of the deleterious effects of OS on fertilization, cleavage, or even implantation (Agarwal et al. 2003; Guerin et al. 2001). Consequently, ICSI manipulation might impair the developmental potential of oocytes after OS injury within the follicular fluid. The results indicate that ROS levels in follicular fluid may have a negative effect on the oocytes and its subsequent development, causing it to be affected by the insemination procedure (Agarwal et al. 2003). Furthermore, it has been recommended that the oocyte-handling time during ART procedures (denuding, ICSI, during media transfer) should be kept at a minimum and that incubation time of oocytes in the culture medium be managed well. This is to enhance oocyte quality and consequently, the success rate of the IVF cycle. For example, exposure time to pro-oxidant media (such as IVF medium) (even before insemination) should be kept at a minimum, as it would be helpful to conserve the quality of oocytes (Martin-Romero et al. 2008).

2.2.4   pH and Temperature

The *in vitro* environment is stressful for gametes and embryos as the temperature and pH tends to fluctuate. The measure of acidity or basicity of a solution is defined as pH, or hydrogen ion concentration. Intracellular pH is a vital aspect of cell homeostasis, and is controlled by membrane potential and osmolarity. The key intracellular processes are highly susceptible to pH, including protein synthesis, metabolism, mitochondrial function and cytoskeletal regulation (Will et al. 2011). In culture media, pH is an important variable that influences motility and sperm binding, oocyte maturation and embryo development, though confounding factors such as bicarbonate and CO_2 levels exist (Will et al. 2011; Bagger et al. 1987).

Animal studies have suggested that even a little increase in external pH (pHe) during minor manipulations outside the incubator can significantly obstruct sperm function (Marquez and Suarez 2007), alter organelle localization(Squirrell et al. 2001; Will et al. 2011), impact the development of mouse blastocyst and hatching, and alter gene expression profiles (Huntriss and Picton 2008). Hamster embryo studies show that even very slight deviations of internal pH (pHi) for short periods of time, from a set point of 7.21 (slightly raised 7.42 or lowered 6.87 pHi), can greatly impact the developing embryo. Philips and his group employed the pH-sensitive fluorescent dye BCECF to evaluate the pHi in human oocytes and demonstrated that the pHi values for oocytes and embryos changes during various developmental stages: such as at GV-intact MI 7.04, MII 7.03 and at Cleavage stage 6.98 pHi (Phillips et al. 2000).

In IVF, the most common buffer used in media is sodium bicarbonate (Will et al. 2011). In addition, the HEPES media buffer has regularly been seen to be a safe and effective buffer for the storing and handling of spermatozoa compared to other types of buffers. CO_2 and pH have an inverse relationship; as CO_2 concentration decreases, pH increases. The media pH may be maintained provided that the levels of CO_2 remain constant in the incubators; nevertheless, this may be a challenge due to frequent openings/closings of incubator doors in order to observe the cell and while performing manipulations at room atmosphere (Will et al. 2011).

In case of procedures performed in room atmosphere, such as gamete collection, ICSI, cryopreservation, and embryo transfer, labs may choose to use handling media with reduced levels of bicarbonate and include another pH buffer to maintain pHe outside the incubator. Sperm medium with added bicarbonate helps with the recovery of motile sperm (Mehta and Sigman 2014). Likewise, temperature is another factor that has an important role in pH and pH buffering. pH and pKa values decrease when the temperature increases (Ferguson et al. 1980). Temperature is measured and maintained by the incubator's thermostat and the incubator's heating system. In an IVF lab, all incubator temperatures are set at 37 °C in order to mimic *in vivo* conditions (Ferguson et al. 1980). Based on the physical properties of these membranes and membrane-associated processes, they may be more sensitive to temperature stress (Davidson and Schiestl 2001). Temperature stress may impair mitochondrial functions and induce oxidative damage, causing lipid peroxidation (Larkindale and Knight 2002).

2.2.5   Centrifugation

Sperm centrifugation is regularly done during semen processing and is a common step used for sperm preparation in ART (Agarwal et al. 2006c). The centrifugation process separates sperm cells from the seminal plasma, and motile sperm from non-motile or dead sperm and cell debris. Some of the commonly used sperm preparation techniques during ART that include the centrifugation step(s) are the double wash sperm swim-up technique, and the double density gradient separation. For example, simple washing (which removes only the seminal plasma) and swim up technique (which involves the further removal cellular debris and non-motile sperm) involve centrifugation at $300–400 \times g$ for about 10 min, while discontinuous (density) gradient involves centrifugation twice, once at $<500 \times g$ for 20 min and then again at $300 \times g$ for 5–10 min. These techniques aid in the selection of sperm with enhanced motility and are more viable. Besides these advantages, the removal of spermatozoa from seminal plasma is important in assisted reproduction, as this step eliminates the seminal plasma that contains leukocytes, a source of ROS. Furthermore, in severely oligospermic semen samples, centrifugation increases the chances for selecting better quality sperm, while in azoospermic semen samples, centrifugation increases the chances of identifying the rare sperm, if any (Sharma et al. 1997).

However, the centrifugation process itself generates ROS (Alvarez et al. 1993; Agarwal et al. 1994; Lampiao et al. 2010). The sperm membrane is mainly made up of polyunsaturated fatty acids (PUFAs) and is therefore particularly susceptible to damage by ROS. High ROS concentrations could lead to lipid peroxidation of sperm plasma membrane, causing the loss of membrane fluidity, which impairs sperm motility. As the concentration of progressively motile sperm is indicative of success of ART outcome/pregnancy, poor sperm motility augurs an unfavourable ART outcome. In addition, the excessive presence of ROS causes DNA damage, which translates to poor embryo development.

Moreover, the g force, time (Shekarriz et al. 1995) and temperature employed during the process of centrifugation, influences the amount of ROS produced and thereby affects the quality of the sperm that has been processed. Centrifugation speeds greater than $500 \times g$ and continuous centrifugation for longer than 5–7 min was found to compromise sperm quality (Sharma et al. 1997). In another study, both 10 and 30 min of sperm centrifugation was found to compromise sperm quality and viability, with the centrifugation time of 30 min being more detrimental to sperm quality compared to 10 min (Lampiao et al. 2010).

Longer centrifugation time increases the temperature during centrifugation, which affects the quality of sperm, despite the initial quality of the sample (Henkel and Schill 2003). The increased temperature during centrifugation also affects sperm motility, which could compromise ART outcome.

Therefore, sperm preparation techniques in ART should ideally exclude the centrifugation step altogether, or should at least avoid the use of prolonged periods of centrifugation (Lampiao et al. 2010). In protocols that require sperm centrifugation, the addition of antioxidants or other ROS scavengers prior to centrifugation may help quench the ROS produced due to the centrifugation process, and yield processed sperm with less damage (Lampiao et al. 2010). For example, pentoxifylline which is added to the sperm preparation to stimulate sperm motility, also quenches the ROS produced by spermatozoa (McKinney et al. 1996).

2.2.6   Culture Media

There are various types of media that may be used to culture human oocytes and pre-implantation embryos during the IVF and ICSI procedures. ART laboratories have a variety of commercially-available culture media to choose from, and these media may be composed of different types of ingredients, depending on the manufacturer. However, the type of media that is eventually used is crucial as culture media used has an important direct bearing on the quality of the embryo produced and subsequently, the success rate of the IVF procedure (Agarwal et al. 2006b). Some media may contain metallic ions such as iron (Fe^{2+}) and copper (Cu^{2+}). These ions can incorporate into the gametes or the developing embryo during culture, leading to the Fenton and Haber-Weiss reactions to occur, which generates ROS (Guerin et al. 2001). To prevent these reactions from taking place and reduce the formation of ROS, common metal chelators such as transferrin and ethylenediamine tetra-acetic acid (EDTA) can be added to the media (Nasr-Esfahani and Johnson 1992; Orsi and Leese 2001).

However, adding supplements to the media could end up increasing the oxidant load. For example, media additives such as serum albumin which contains elevated levels of amine oxidase leads to a higher generation of hydrogen peroxide, a form of ROS (Shannon 1978; du Plessis et al. 2008). The rate with which the culture media generates ROS varies according to its composition (Jana et al. 2010). OS within the culture media could partially deplete the GSH content in the oocyte, which could disrupt oocyte fertilization and viability (Martin-Romero et al. 2008).

Supplementation of culture media with antioxidants such as ascorbic acid, alpha-tocopherol, taurine, hypotaurine, isoflavones reduced the risk of OS and its subsequent adverse effects on the gamete (Sikka 2004; Alvarez and Storey 1983). Lipid peroxidation due to the presence of ROS can be deterred using vitamin E supplementation (Jain et al. 2000). In mouse embryos, the addition of thioredoxin to the media reduced apoptosis rates and enhanced hatching rates, while in porcine embryos, supplementation of media with glutathione and thioredoxin reduced the redox status (Ozawa et al. 2006).

In the *in vitro* fertilization procedure, sperm from the male partner interacts with the oocyte from the female partner in culture media, leading to fertilization. In ICSI, sperm prepared in culture media is injected directly into the cytoplasm of the oocyte. As the sperm is injected into the oocyte, there is an additional risk of transferring some of the ROS that is present in the culture media along with the sperm into the oocyte—which would have further detrimental effects on the oocyte's DNA material (Shen et al. 2003).

The consequences of ROS on gametes and early embryos have been experimentally-investigated, and multiple authors have reported increased levels of intracellular ROS during the various stages of embryo development (Hu et al. 2001; Bedaiwy et al. 2004).

Bedaiwy and his group examined the association of ROS levels in the culture media on the first day (day 1 ROS) after ICSI (Bedaiwy et al. 2004). During this experiment, fertilization as well as early cultures were done in human tubal fluid supplemented with 5 % serum substitute and levels of ROS were monitored by a chemiluminescence method using a luminol probe. The results illustrated that high ROS levels in culture media on day 1 caused low blastocyst rate, low fertilization rate, low cleavage rate, and high embryonic fragmentation in ICSI cases but not in those of conventional IVF. However, lower pregnancy rates were observed in both IVF and ICSI cycles with high day 1 ROS levels in the culture media. The same group followed up the experiment on examining ROS levels in day 3 culture media to the outcome of ICSI (Bedaiwy et al. 2010). The results confirmed that increased levels of ROS production in day 3 embryo culture media have detrimental effects on the embryo growth parameters as well as clinical pregnancy rates.

From the studies discussed, we can conclude that in order to improve sperm quality, embryo development as well as clinical pregnancy rates, short co-incubation time for gametes should be clinically applied in an attempt to improve ART outcome. Furthermore, supplementing of the culture media with required antioxidants to help scavenge excessive production of ROS in culture media is beneficial.

In addition, the design of the culture media used should be such that the media takes into consideration the role of ROS generation and eliminate their contribution without changing the metabolic requirements of gametes and embryos.

2.2.7   Oxygen Concentration

Gametes/embryos are exposed to oxygen tension during procedures in assisted reproduction such as insemination, fertilization and embryo growth (Catt and Henman 2000). The development of the human pre-implantation embryo *in vitro* is influenced by the atmospheric oxygen partial pressure and dissolved oxygen concentration in the culture medium. In ART laboratories, cells are commonly cultured *in vitro* in an atmospheric oxygen concentration of 20 % (160 mmHg) or an incubator environment of ~20 % oxygen and 5 % carbon dioxide. However, under physiological conditions *in vivo* in the oviduct and uterus, embryos are exposed to much lower oxygen concentrations of 2–8 % or 19–70 mmHg (Calzi et al. 2012). At 37 °C, the oxygen concentration in the medium equilibrated with atmospheric oxygen was found to be 20-times higher than physiological intracellular oxygen concentration (Jones 1985). Thus, the use of dissolved oxygen levels in the culture media *in vitro* at ~5 %, which is adopted by ART laboratories nowadays, is more comparable to the oxygen concentration at the tissue level *in vivo*.

Oxygen plays an essential role in cell growth and differentiation, but in an IVF setting, the presence of high concentrations of oxygen during incubation activates various cellular oxidase enzymes. This in turn increases the generation of ROS—leading to a state of OS (Cohen et al. 1997). Thus, the gametes and media in the ART laboratory setting should be protected from exposure to high partial pressures of oxygen in order to minimize the production of ROS during IVF procedures. When compared to embryos cultured under atmospheric oxygen conditions, those cultured at 5 % oxygen yielded better developed embryos and higher pregnancy rates (even among the poor responders), during IVF and ICSI cycles (Kovacic and Vlaisavljevic 2008; Kovacic et al. 2010). Interestingly a meta-analysis of seven randomized controlled trials (RCTs) comparing the effects of oocyte/embryo culturing at low (~5 %) and atmospheric (~20 %) oxygen concentration on assisted reproduction outcomes such as fertilisation, implantation and ongoing pregnancy rates showed no significant difference between the two groups (Gomes-Sobrinho et al. 2011). Embryos transferred on days 2 or 3 had similar implantation rates regardless of the oxygen tension used (~5 % vs. ~20 %) during culture, but for embryos transferred on day 5 or 6 (blastocyst stage), the implantation rates of embryos cultured at ~5 % oxygen were significantly higher than embryos cultured at ~20 % oxygen. However, the ongoing pregnancy rates did not differ significantly despite the different oxygen tensions used (~5 % or ~20 %) or the day of transfer (days 2/3 or days 5/6) (Gomes-Sobrinho et al. 2011).

Results of a recent Cochrane systematic review (involving 7 studies, 2,422 participants) and meta-analysis (involving 4 studies, 1,382 participants) showed that embryos cultured in low (5 %) oxygen concentrations developed better and were therefore of improved quality. This resulted in higher ongoing and clinical pregnancies rates and improved live birth rates. Thus compared to embryo culture in atmospheric (~20 %) oxygen concentrations, the culture of preimplantation embryos under low (~5 %) oxygen concentrations improves IVF/ICSI success rates and results in the birth of healthier babies (Bontekoe et al. 2012).

To strengthen these positive results, larger, well-designed randomized controlled trials are required. Results from these type of studies would provide more conclusive evidence on the impact of low oxygen culture on IVF outcome. In general, incubation of spermatozoa is usually done at 37 °C in an atmosphere of 5 % CO_2 for at least 1–2 h before ICSI or conventional insemination. In addition, these types of preparation and incubation conditions could affect the DNA integrity of ejaculated human spermatozoa.

Chapter 3
Antioxidant Strategies

As has been portrayed in the previous section, there is a growing body of evidence to signify that multiple *in vivo* and *in vitro* factors could encroach on an ART setting, leading to an increase in OS and ART outcome. Therefore in ART laboratories, it is indispensable to scrutinize approaches that will help increase the success rate of the procedure by limiting the potentially negative effects on successful ART outcomes (Sikka 2004).

Elevated ROS levels need to be quelled as much as possible to suppress its detrimental effects on the ART procedure. The effects of high levels of ROS on the human gametes, fertilization and embryo are summarized in Fig. 3.1. ROS are neutralized by an intricate defence system consisting of enzymatic antioxidants and non-enzymatic antioxidants. Preferably, an antioxidant should reach high enough concentrations and supply the deficient vital elements required for the process to occur. Furthermore, the antioxidant should have the aptitude to enhance the scavenging capacity and maintain a physiological amount of ROS, without suppressing it entirely (Zini et al. 2009). Therefore, the first step to take would be to establish the fundamental cause of the ROS/antioxidant imbalance and treat it (Agarwal et al. 2004).

As discussed earlier, the body contains many natural antioxidant defence systems *in vivo*. These mechanisms become eradicated in the *in vitro* state. Thus, in order to aid in the management of OS during ART procedures, the potential patients could be provided with oral antioxidant supplements prior to the ART procedure. Another option would be to add various dosages of antioxidants directly into the media during the ART procedure in order to reduce the development and effects of OS.

Oral antioxidants have been reported to improve the quality of gametes prior to the IVF procedure. It has also been shown to enhance fertilization and pregnancy rates. In the following section, we will discuss several studies that support the use of specific enzymatic antioxidants including SOD, glutathione reductase (GSH) and catalase; some non-enzymatic oxidants including vitamins C and E, folic acid, L-Carnitine, Coenzyme Q_{10} and melatonin as well as combined antioxidant strategies (Lampiao 2012). Table 3.1 summarizes the mechanisms of action and effects of

© The Author 2014   23
A. Agarwal et al., *Strategies to Ameliorate Oxidative Stress During Assisted Reproduction*, SpringerBriefs in Reproductive Biology, DOI 10.1007/978-3-319-10259-7_3

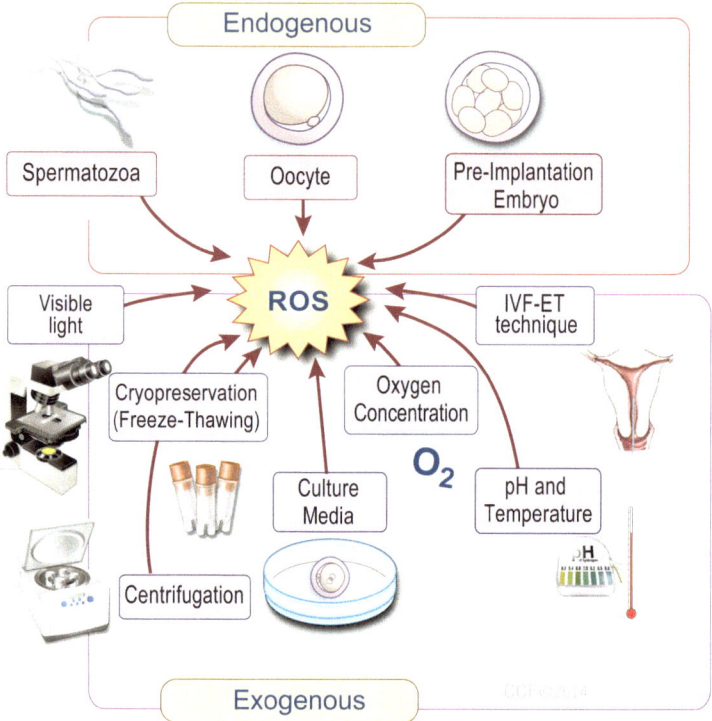

Fig. 3.1 Effects of pathological levels of reactive oxygen species in assisted reproductive technology

antioxidants that is of importance and that are frequently-used clinically. Despite non-enzymatic antioxidant supplementation being more common than enzymatic antioxidants, both types of antioxidants have a significant role to play in minimizing of OS. In the coming section, we will also indicate the exact dosage information and methods of how each antioxidant treatment works, as well as provide recommendation for its use in ART, based on the outcome of the studies discussed.

3.1   Role of Enzymatic Antioxidants in ART

In vivo, the superoxides are released by oxidative phosphorylation and other processes. They are transformed initially to hydrogen peroxide H_2O_2 and then reduced to water. This detoxification pathway has multiple enzymes—superoxide dismutases begin the catalyses, then catalases and various peroxidases remove H_2O_2. The contributions of these enzymes to antioxidant defences are difficult to separate from one another. The significance of enzymatic antioxidants in males has been reported in various studies, and will be discussed in the following section.

Table 3.1 Role of antioxidants in ameliorating the effects of oxidative stress in assisted reproduction

Antioxidant(s)	Study	Treatment dosage (per treatment or day)	Treatment details	Parameters improved
SOD	Rossi et al. (2001)	SOD (100 U/mL) and catalase (100 U/mL)	Added to fresh semen before freezing	• Semen parameter, especially progressive motility) (SOD and catalase)—due to their combined and simultaneous action on superoxide anion and hydrogen peroxide
Catalase	Li et al. (2010)	Ascorbate (300 or 600 μM) and catalase (200 or 400 IU/mL)	Added to semen before freezing	• Reduced ROS levels and ROS-induced damages in post-thaw spermatozoa [ascorbate (300 μM) and catalase (200 and 400 IU/L)]
Catalase	Chi et al. (2008)	Catalase (1, 10, 100 U/mL) or EDTA (1, 10, 100 μM/mL)	Added to medium used during sperm wash	• Sperm motility (10 μM/mL EDTA) • Acrosome reaction rate of the spermatozoa (catalase) • Decreased DNA fragmentation rate of the spermatozoa (EDTA and catalase)
Vitamin E	Kalthur et al. (2011)	Vitamin E 5 mM	Added to cryomedia prior to freeze-thaw	• Post-thaw motility • DNA integrity
Vitamin E	Taylor et al. (2009)	Vitamin E 100 or 200 μmol	Added to cryomedia	• Post-thaw motility
Vitamin E	Cicek et al. (2012)	Vitamin E 400 IU orally to women with unexplained infertility undergoing ovarian stimulation and IUI	Between day 3 and 5 of menstrual cycle until hCG injection	• Endometrial thickness on hCG day
Vitamin E + selenium	Moslemi and Tavanbakhsh (2011)	Vitamin E 400 IU+Selenium 200 μg orally in men with idiopathic asthenozoospermia	100 days	• Sperm motility • Sperm morphology • Spontaneous pregnancy rates
Vitamin C	Akmal et al. (2006)	2,000 mg vitamin C in oligozoospermic men	2 months	• Mean sperm count • Sperm motility • Sperm morphology

(continued)

Table 3.1 (continued)

Antioxidant(s)	Study	Treatment dosage (per treatment or day)	Treatment details	Parameters improved
Vitamin C + vitamin E	Greco et al. (2005b)	Oral 1 g vitamin C and 1 g vitamin E in men with elevated sperm DNA fragmentation (15 %) and prior failed ICSI attempt	2 months	• Reduced DNA damaged sperm • Implantation rates • Implantation rates • Clinical pregnancy rates
Vitamin C	Henmi et al. (2003)	Oral ascorbic acid 750 mg in women with luteal phase defects	1st day of the third cycle until a positive urinary pregnancy test	• Clinical pregnancy rates
Vitamin C	Crha et al. (2003)	Vitamin C 500 mg in women undergoing IVF-ET	Gradual release over 8–12 h	• Improved pregnancy rates
Vitamin C	Branco et al. (2010)	Ascorbic acid 10 mM	Semen before freezing	• Reduced DNA damage
Melatonin	Eryilmaz et al. (2011)	Melatonin 3 mg in IVF-ET patients	3rd to the 5th day of the menstrual cycle until hCG	• Mean number of retrieved oocytes • Mean number of MII oocyte counts • G1 embryo ratio
Melatonin	Tamura et al. (2008)	Melatonin 3 mg in women with prior IVF-ET failure	Between the 3rd and the 5th day of the menstrual cycle until hCG injection	• Oocyte quality • Fertilization rates
Melatonin	Unfer et al. (2011)	Myo-inositol 4 g and melatonin 3 mg and folic acid 400 mcg in women with failed IVF cycle	3 months	• Number of MII oocytes retrieved • Total and top-quality embryos transferred • Fertilization rate
Coenzyme Q_{10}	Nadjarzadeh et al. (2011)	CoQ_{10} 200 mg in iOAT patients	12 weeks	• Increased TAC • Reduced LPO
Coenzyme Q_{10}	Safarinejad (2012)	CoQ_{10} 600 mg in iOAT patients	12 months	• Sperm quality (concentration, progressive motility, morphology) • Pregnancy rates

L-Carnitine	Lenzi et al. (2003)	L-Carnitine 2 g	2 months	• Semen quality
L-Carnitine and acetyl L-Carnitine	Cavallini et al. (2004)	L-Carnitine 2 g and acetyl-L-Carnitine 1 g	6 months	• Semen quality • Pregnancy rates
L-Carnitine with Vitamin E	Wang et al. (2010)	L-Carnitine 2 g and vitamin E	3 months	• Percentage of forward motile sperm after the treatment • Pregnancy rate
L-Carnitine and acetyl L-Carnitine	Vicari and Calogero (2001)	L-Carnitine 2 g and acetyl-L-Carnitine 1 g	3 months	• Sperm forward motility • Sperm viability • Reduced ROS production
L-Carnitine	Khademi et al. (2005)	L-Carnitine 3 g in men with idiopathic sperm abnormalities	3 months	• Percentile of motile and grade A sperm • Percentile of normal-shaped sperms decreased significantly
L-Carnitine	Abdelrazik et al. (2009)	L-Carnitine 0.3 and 0.6 mg/mL	Added to cryomedia	• Blastocyst development rate • Reduced DNA damage
Folic acid and zinc sulphate	Wong et al. (2002)	Folic acid 5 mg and zinc sulphate 66 mg in subfertile men	26 weeks	• Folate concentrations in seminal plasma • Total normal sperm count
Myo-inositol and folic acid	Papaleo et al. (2009)	Myo-inositol 4 g and 400 μg folic acid twice a day	Continuous from the day of GnRH administration	• Reduced mean number of immature oocytes at pick up • Reduced mean number of degenerated oocytes at pick up • Number of retrieved oocytes maintained
Myo-inositol and folic acid	Ciotta et al. (2011)	Myo-inositol 4 g and 400 μg folic acid twice a day	3 months	• Number of oocytes recovered at pick-up • Reduced number of immature oocytes

(continued)

Table 3.1 (continued)

Antioxidant(s)	Study	Treatment dosage (per treatment or day)	Treatment details	Parameters improved
Combination of antioxidants	Wirleitner et al. (2012)	FertilovitRMplus twice daily	2 months	• Sperm motility • Reduced percentage of immotile sperm cells • Total sperm count • Percentage of class I sperm
Combination of antioxidants	Tunc et al. (2009)	1 Menevit capsule	3 months	• Pregnancy outcome
Combination of antioxidants	Tremellen et al. (2007)	Menevit	3 months	• Viable pregnancy rate
Combination of antioxidants	Omu et al. (2008)	Zinc 5 mg, Vitamin E + zinc 10 mg and Zinc + Vitamin E + C 200 mg	3 months	• Sperm parameters
Combination of antioxidants	Rizzo et al. (2010)	Myo-inositol 4 g and folic acid 200 mg and melatonin 3 mg	Continuous from the day of GnRH administration	• Oocyte quality • Number of morphologically mature oocytes at ovum pick up
Combination of antioxidants	Dashti et al. (2013)	Tamoxifen 20 mg, vitamin E 400 IU, zinc 30 mg and selenium 200 mg	3 months prior to IUI	• Sperm concentration, motility, forward progression and the percentage normal forms

3.1.1 Superoxide Dismutases

Within the cell, superoxide dismutases (SODs) constitute the first line of defence against ROS (Alscher et al. 2002). They catalyze the breakdown of the O_2^- anion into oxygen (O_2) and water (H_2O) (Hammadeh et al. 2008). There are three isoforms of SODs, the intracellular mitochondrial isoform, manganese SOD (Mn SOD), and the cytoplasmic isoform, Copper/Zinc SOD (Cu-Zn SOD), along with the extracellular SOD (Fe SOD). In males, this antioxidant group is naturally found in the seminal plasma (Matos et al. 2009).

Hammadeh et al. (2008) observed the effects of various enzymatic antioxidants in the seminal plasma of males undergoing IVF and ICSI treatments. A negative correlation was seen between seminal plasma ROS concentration in and membrane integrity, normal morphology as well as fertilization rate. Furthermore a positive association was found between SOD levels and normal sperm morphology; thus concluding that enzymatic antioxidants definitely do play a positive role in male infertility.

The activity and relationship of SOD with MDA (as a marker of lipid peroxidation), in normozoospermic and asthenozoospermic males has also been tested (Tavilani et al. 2008). The findings indicate a protective role for seminal plasma antioxidant enzymes against lipid peroxidation of spermatozoa in normozoospermic samples. The relationship between seminal antioxidant capacity and OS markers, along with semen quality has been investigated in various studies. Studies that looked at the sperm count and progressive motility, found a significant positive correlation with SOD during IVF (Shiva et al. 2011; Marzec-Wroblewska et al. 2011; Atig et al. 2012). Results from these studies suggest that decreased antioxidant enzymes and increased OS may be the cause of higher risk of deteriorating semen quality. Highly significant and positive correlations were also found between seminal SOD activity and semen parameters such as sperm concentration and overall motility in the normozoospermic group, which are regarded as the most important criteria for normal fertilizing ability of the spermatozoa.

In women with tubal factor infertility, the association between SOD, MDA and the incidence of apoptosis of granulosa cells in follicular fluid was investigated. It was demonstrated that there was a significant negative correlation between MDA and SOD levels and significant positive correlation between MDA levels and the prevalence of apoptosis (Liu and Li 2010). It was also noted that patients who had undergone a failed cycle of IVF-embryo transfer (IVF-ET) had a propensity to have lower levels of SOD present in their granulosa cells. It was furthermore evident that these patients had embryos of poorer quality compared to the group of patients who were able to achieve pregnancy. In these patients, SOD has been reported to increase the proportions of zygotes that undergo at least one round of cleavage while improving cleavage past the two-cell stage and overall blastocyst development (Liu and Li 2010).

3.1.2 Catalase

Human seminal plasma contains catalase which is principally derived from the prostate gland. In females, it is localised in the corpus luteum and tubal fluid (Rakhit et al. 2013). This enzyme catalyzes the breakdown of H_2O_2 to H_2O and O_2. In males, catalase is known to improve the motility, vitality and DNA integrity of cryopreserved human spermatozoa (Moubasher et al. 2013). In their study, Moubasher's group added 200 μL/mL of catalase to semen during cryopreservation, and the sample was then cryopreserved for 24 h. The post-thaw results showed a momentous increase in the percentage of progressively motile and viable spermatozoa, while the % of DNA damage significantly decreased in samples supplemented with catalase compared to samples without catalase (either fresh or processed). The outcome of this study was in agreement with another study which evaluated the consequence of both SOD and catalase supplementation in human semen undergoing cryopreservation (Rossi et al. 2001). From the results, it was evident that the addition of SOD (100 U/mL) and catalase (100 U/mL) to semen prior to cryopreservation improved the percentage recovery of progressively motile spermatozoa after thawing. These results suggested that, SOD and catalase supplementation can contribute significantly to the prevention of sperm membrane lipid peroxidation by ROS and thus allow for good sperm parameter recovery after freeze-thawing procedures.

The protective effects of ascorbate and catalase on cryopreserved spermatozoa has also been investigated (Li et al. 2010). Ascorbate (300 or 600 μM) and catalase (200 or 400 IU/mL) were added to semen samples. Sperm viability, motility, and mitochondrial membrane potential was a significantly lower in comparison to that of fresh spermatozoa. Moreover, there was an increase in apoptosis [positive for annexin V and negative for propidium iodide Ann(+)/PI(−)] and DNA damage (olive tail movement (OTM)) in the cryopreserved spermatozoa . It was observed that ascorbate and catalase reduced the ROS levels in post thaw spermatozoa significantly. Therefore appropriate ascorbate or catalase supplementation to the cryoprotective medium can control ROS levels and minimize the resultant cryodamage.

Different concentrations of catalase (100, 10, 1 U/mL, Sigma) and ethylenediaminetetraacetic acid (EDTA) (a metal chelating agent) were added to spermatozoa while washed with Ham's F-10 media (Chi et al. 2008). The researchers indicated that the addition of EDTA to sperm preparation medium improved sperm motility compared to control group, EDTA group at various concentrations and catalase group. Interestingly, catalase significantly improved the acrosome reaction rate of the spermatozoa. EDTA and catalase both significantly lowered the sperm DNA fragmentation rate.

A recent study demonstrated the influence of seminal catalase activity on IVF pregnancy rates (Zelen et al. 2010). Semen samples were collected from men undergoing IVF and ICSI. Conventional semen parameters, sperm DNA fragmentation index (%DFI), high DNA stainability (%HDS) and seminal catalase-like activity were measured and the results correlated with outcomes of IVF and ICSI. It was discovered that catalase activity was positively correlated with IVF pregnancy rate but not with ICSI pregnancy rate. The data suggested that couples in whom the men have low seminal catalase activity (i.e., inadequate protection from hydrogen peroxide) have a higher risk of post-fertilization failure during IVF cycles and it is therefore advisable that ICSI should be used as alternative treatment option.

3.1.3   Glutathione Peroxidase

Glutathione peroxidase (GPX) governs one of the most important systems in the exclusion of ROS in the seminal plasma. In the female, it is located within the glandular epithelium of the endometrium. GPX catalyse the reduction of organic and inorganic hydroperoxides, using reduced glutathione as an electron donor. Few studies have documented the role of GPX activity in human seminal plasma. A recent investigation studied the correlation between GPX activity levels in human seminal plasma with standard semen parameters and spermatozoa fertilization potential, in terms of fertilization and pregnancy rates in IVF (Crisol et al. 2012). They found that seminal plasma GPX activity was significantly lower in patients with abnormal sperm compared to normozoospermic individuals. These results were in agreement with an earlier study, that determined the expression and enzymatic activity of GPX-1 and GPX-4, concentrations in spermatozoa from fertile and infertile men (Garrido et al. 2004). In Garrido's study, men with poor sperm morphology had significantly lower levels of GPX activity. They suggested that sperm GSH system components in the cell, GPX-4 and GSH, vary in infertile men. These changes may be associated with sperm morphology. From these results, we can conclude that GPX activity in human semen is critical for normal sperm function and morphology (Garrido et al. 2004).

3.2   Role of Non-enzymatic Antioxidants in ART

Non-enzymatic antioxidants, also known as synthetic antioxidants or dietary supplements (Agarwal et al. 2005b) involve a multitude of low molecular mass ROS scavengers such as vitamin E, C, melatonin and many other naturally-occurring antioxidants (Gharagozloo and Aitken 2011). Based on the credence of such scientific evidence (Table 3.1), clinical experiments have been carried out to ascertain the beneficial effects of non-enzymatic antioxidants in improving ART outcome. Couples undergoing ART infertility may benefit from the use of these non-enzymatic antioxidants in order to improve OS situations encountered during IVF/ICSI. These antioxidants have been evaluated clinically either individually or in combination. The study trials that are elaborated upon in the following sections include studies that have been conducted in various population sizes, types, doses as well as duration of antioxidant therapy.

3.2.1 Vitamin E

Vitamin E refers to a group of eight fat-soluble compounds that consist of both tocopherols and tocotrienols. Alpha-tocopherol is one of the most important lipid-soluble antioxidant molecules as well as the most biologically active form of vitamin E, located mainly in the cell membranes, (the second-most common form of vitamin E in the diet). It acts by breaking pathological ROS-induced chain reactions. It confers its protective effects by shielding sperm membrane components from oxidative stress damage (Mora-Esteves and Shin 2013). Various studies discussed in this section evaluated the effects of vitamin E alone or in combination with other vitamins.

In clinical trials, vitamin E supplementation has been found to increase semen parameters and fertilization rates possibly by reducing oxidative damage and lipid peroxidation potential during IVF/ICSI treatment (Comhaire et al. 2000; Geva et al. 1996). A randomized controlled trial was set up to investigate the action of vitamin E in men with idiopathic infertility (Cicek et al. 2012). In this study, the group analysed vitamin E effect in patients with unexplained infertility undergoing intrauterine insemination (IUI). It was hypothesized that clinical outcomes of IUI cycles may improve with vitamin E administration. 400 IU/day of vitamin E was administered beginning from between the 3rd to the 5th day of the menstrual cycle, until the human chorionic gonadotropin (hCG) injection day. The results demonstrated that vitamin E supplementation in unexplained infertile patients had beneficial effects in improving the endometrial thickness during IUI cycles. Furthermore, higher implantation and ongoing pregnancy rates were observed in the vitamin E-administered group, and it was concluded that these improvements may be a result of the improving antioxidant effect of vitamin E on the endometrial receptivity.

The effect of addition of vitamin E to cryoprotective media prior to sperm cryopreservation on the post-thaw motility and DNA integrity of normozoospermic and asthenozoospermic semen samples was also investigated (Kalthur et al. 2011). The group noted that supplementation of 5 mM vitamin E to cryomedia significantly improves the post-thaw motility and DNA integrity in normozoospermic and asthenozoospermic semen samples prior to ICSI. The results of this study were in conformity with a previous study that investigated whether the addition of an antioxidant to cryopreservation medium could improve the post-thaw integrity of cryopreserved human spermatozoa, particularly from men with abnormal semen parameters. The cryopreservation medium contained either 100 or 200 μmol of vitamin E. The study findings report of a significant improvement of post-thaw motility (Taylor et al. 2009).

The efficacy of vitamin E in combination with selenium (Se) for improving semen parameters and pregnancy rates were also investigated in various studies. Se, an essential trace element occurring in organic and inorganic forms, is important for reproductive functions such as testosterone metabolism and is a constituent of the sperm capsule selenoprotein (Mora-Esteves and Shin 2013).

The effects of vitamin E and selenium supplementation on lipid peroxidation and on sperm parameters were inspected by one of the earlier groups (Keskes-Ammar et al. 2003). 400 mg of vitamin E and 225 μg of selenium was provided daily to male patients for a period of 3 months. Findings of the study include a significant decrease in MDA levels and an improvement in sperm motility after treatment. These results were also in agreement with a recent study that examined the efficacy of vitamin E in combination with selenium for improving semen parameters and pregnancy rates (Moslemi and Tavanbakhsh 2011). Patients were assigned to fixed-dose treatment with oral 200 µg selenium tablet (L-selenomethionine) daily in combination with 400 IU of synthetic vitamin E (α-tocopherol) for 100 days. Total enhancement in sperm motility and normal morphology, as well as in spontaneous pregnancy were observed when compared with the controls. Overall, a beneficial and protective effect of Vitamin E on semen quality was suggested, particularly sperm motility. Furthermore, the authors advocated supplementation of vitamin E and selenium for the treatment of idiopathic male infertility diagnosed with asthenoteratozoospermia or asthenozoospermia in semen analysis.

3.2.2   Vitamin C

Vitamin C (L-ascorbic acid, ascorbate) has long been allied with fertility. This vitamin is a crucial water-soluble micronutrient obligatory for an array of biological functions. It is an unstable, easily-oxidized acid and can be damaged by oxygen, alkali and high temperature. In males, it is considered a major antioxidant in the testes and highly concentrated in seminal plasma where it has also been reported to contribute up to 65 % of its total chain breaking antioxidant capacity (Augustine et al. 2005).

Several studies have reported a positive relationship between seminal vitamin C and sperm quality as studied in both fertile and infertile men. Infertile patients were reported to have significantly lower vitamin C levels in their seminal plasma compared to fertile men (Colagar and Marzony 2009). Their findings suggested that idiopathic infertile men have significantly lower levels of ascorbic acid in their seminal plasma than fertile men. Thus, the authors recommended analysis of ascorbic acid levels in seminal plasma of patients with idiopathic infertility as ascorbic acid supplementation could help improve semen parameters in these patients. A study was conducted to observe if the increased incidence of DNA fragmentation in ejaculated spermatozoa during the IVF treatment could be reduced by addition of vitamins C and E. The study results showed that a 2 month daily oral supplementation of vitamins C (1 g) and vitamin E (1 g) reduced the percentage of DNA-fragmented spermatozoa (Greco et al. 2005a). For such spermatozoa DNA fragmentation cases, The same group also discovered that oral antioxidant treatment improves ICSI outcome (Greco et al. 2005b).

A treatment protocol consisting of daily oral antioxidant intake (vitamins C and E, beta-carotene, zinc and selenium) caused a marked improvement in clinical pregnancy and implantation rates in patients with >15 % sperm DNA fragmentation index. Menezo and his group also suggested that this oral antioxidant treatment appears to improve ICSI outcomes in those patients with sperm DNA damage, however, the degree of sperm decondensation of these patients should also be considered (Menezo et al. 2007). The efficacy of vitamin C supplementation in cryomedium reported improved sperm motility preservation post thaw. The experiments involved addition of 10 mM of ascorbic acid to semen before cryopreservation. It also reduced DNA damage and preserved the integrity of spermatozoa in infertile men (Jenkins et al. 2011; Branco et al. 2010).

The effect of oral supplementation of vitamin C on various semen parameters in oligospermic, infertile men during IVF/ICSI treatment was monitored (Akmal et al. 2006). Various semen parameters, including sperm motility, sperm count, and sperm morphology, were studied before and after the vitamin C treatment. These patients received in an open trial of 1,000 mg of vitamin C twice daily for a maximum of 2 months. Results showed that the mean sperm count was increased. The mean sperm motility was increased significantly and mean sperms with normal morphology increased significantly as well suggesting that vitamin C supplementation in infertile men might help improve these parameters. The impact of vitamin C supplementation was also observed in females undergoing IVF-ET procedure The influence of vitamin C supplementation on the outcome of infertility treatment 76 women (38 of them smokers, the remaining 38 were non-smokers) was studied (Crha et al. 2003). The women were administered vitamin C in daily doses of 500 mg. The results proved that vitamin supplementation had a greater impact on the number of pregnancies in the non-smokers' group. The pregnancy rate was significantly higher in non-smoking women than in smokers which appears to be a reason for asking women to cease smoking prior to infertility treatment. Further, on the addition of vitamin C to culture media, increase sperm motility and viability along with decreases in MDA levels were noted (Mostafa et al. 2006).

3.2.3 Melatonin

Melatonin (N-acetyl-5-methoxytryptamine) is a multi-functional and universal antioxidant (Ronnberg et al. 1990), secreted during the dark hours at night by pineal gland, and it regulates a variety of significant central and peripheral actions related to circadian rhythms and reproduction (Ronnberg et al. 1990). Melatonin has been demonstrated as a powerful direct scavenger of free radicals in a growing number of studies, and is also known for its multifunctional and universal antioxidant properties (Eryilmaz et al. 2011). Melatonin regulates ovarian function by the instruction of gonadotropin discharge in the hypothalamus-pituitary gland axis via its specific receptors (Eryilmaz et al. 2011).

A growing number of studies have demonstrated the role of melatonin in reproduction. Human follicular fluids contain higher melatonin concentrations than plasma which increases depending on the follicular growth (Nakamura et al. 2003; Ronnberg et al. 1990). The relationship between OS, oocyte quality and melatonin was also investigated in patients undergoing IVF-ET (Tamura et al. 2008). 3 mg/day melatonin was administered in women from the fifth day of the previous menstrual cycle until the day of oocyte retrieval. Intrafollicular concentrations of 8-OHdG and hexanoyl-lysine adduct were significantly reduced by these antioxidant treatments. The fertilization rate was improved by melatonin treatment and suggested that melatonin protects oocytes from OS and is likely to improve the oocyte quality and fertilization rates. The same treatment protocol was followed up in another study (Eryilmaz et al. 2011). The results illustrated that the mean number of the retrieved oocytes, the mean MII oocyte counts, the G1 embryo ratio were significantly higher in the melatonin-administered group than that of the non-melatonin-administered group.

The efficacy of a treatment with myo-inositol plus folic acid plus melatonin compared with myo-inositol plus folic acid alone on oocyte quality was investigated in women who underwent IVF (Rizzo et al. 2010). Patients were assigned to obtain either 2 g myo-inositol twice a day combined with 200 mg folic acid and 3 mg melatonin. The data showed improvement in oocyte quality, number of morphologically mature oocytes at ovum pick up. Results from the study revealed significantly top-quality embryos in patients treated with melatonin. In the following year, another group conducted a study to evaluate the pregnancy outcomes after the administration of myo-inositol combined with melatonin in women who failed to conceive in previous IVF cycles due to poor oocyte quality (Unfer et al. 2011). Women were treated with 4 g/day myo-inositol and 3 mg/day melatonin for 3 months and then underwent a new IVF cycle. After treatment, the number of mature oocytes, the fertilization rate, the number of both, total and top-quality embryos transferred were statistically higher compared to the previous IVF cycle.

The results were in agreement with another group that evaluated the effect of melatonin alone for improving oocyte quality in patients with previous failed IVF treatment (Batioglu et al. 2012). 1 or 3 mg tablets of melatonin were taken orally by the patients from the fifth day of the previous menstrual cycle to the day they were injected with HCG. The results revealed percentage of mature oocytes (M2/oocytes retrieved) was significantly different in melatonin-treated group. The mean number of class 1 embryos resulted higher in the melatonin-treated group as well as clinical pregnancy rate in women undergoing IVF or ICSI was higher.

In males, melatonin is localised in seminal fluid and membrane receptors of spermatozoa. The *in vitro* effects of melatonin on human sperm motility, concentration and motility was evaluated (Ortiz et al. 2011). The data discovered urinary 6-sulfatoxy melatonin and total antioxidant capacity levels were positively correlated with sperm concentration, motility and morphology, as well as negatively correlated with the number of round cells. Additionally, 30-min exposure of sperm to 1 mm melatonin improved the percentage of motile and progressively motile cells.

3.2.4 Vitamin B: Folic Acid

An additional empirically-used and an inspected agent in infertility is folic acid. They are group B vitamins implicated in the one-carbon metabolism requisite for purine and pyrimidine synthesis and eventually affecting DNA synthesis and integrity (Joshi et al. 2001). They are also involved in re-methylation of homocysteine (HCY) into methionine which is further metabolized into *S*-adenosylmethionine, the universal methyl donor for transmethylation of DNA. Homocysteine being a methionine catabolite and its concentration in follicular fluid, is negatively related to oocyte maturity (Papaleo et al. 2011).

As folate plays a key role in germ cell development, consequently it is noticeable that folate is important for reproduction (Joshi et al. 2001; Ebisch et al. 2007). In order to justify, it was established that seminal plasma total folate concentrations reflect folate nutriture and that it may be important for male reproductive function. It was demonstrated that total seminal plasma folate concentrations were on average 1.5 times higher than blood plasma folate concentrations in men (Wallock et al. 2001).

It was experimented that 5 mg folic acid administration to subfertile and fertile men for 26 weeks resulted in a significant increase of folate concentrations in seminal plasma. Although no effect of this involvement was observed on sperm count or motility of spermatozoa, however there was a 74 % increase in total normal sperm count after a combination of folic acid and zinc sulphate in both subfertile and fertile me (Wong et al. 2002).

In females, a study was conducted to evaluate the association between follicular fluid homocysteine concentration and the degree of maturity of oocyte (Szymanski and Kazdepka-Zieminska 2003). 40 patients were qualified for IVF-ET and underwent folic acid supplementation. The folic acid concentration recovered from follicles was measured using microparticle enzyme immunoassay (MEIA). The rationale of the study was to establish dependencies between the concentration of homocysteine in follicular fluid and the quality of oocytes and it was discovered that in a group of women with folic supplementation and lower homocysteine concentration the percentage of oocytes in first and second degree of maturity was higher and of better quality. The effects of myo-inositol and folic acid on oocyte quality in PCOS patients undergoing ICSI cycles were determined (Papaleo et al. 2009). The participants received myo-inositol combined with 2 g folic acid twice a day. The results of their experiment revealed that in patients with PCOS, treatment with myo-inositol and folic acid reduces germinal vesicles and degenerated oocytes at ovum pick-up compared to control. The results were in agreement with the outcome of another study that determined the effects of myo-inositol plus folic acid on oocyte's quality in PCOS patients (Ciotta et al. 2011). 2 g of myo-inositol plus 200 mg of folic acid were taken by the participants twice a day, continuously for 3 months. The results at the end of treatment revealed that the number of oocytes recovered at the time of pick-up were significantly greater as compared to control. Significantly reduced number of immature oocytes (vesicles germ and degenerated oocytes) was also investigated. Data from both the studies suggest that this treatment combination may be useful in the management of PCOS patients.

3.2.5   Coenzyme Q_{10}

Coenzyme Q_{10} (CoQ_{10}) also commonly called ubiquinones mostly produced in mitochondria of cells and present in human seminal fluid, exerts important metabolic and antioxidant functions and shows direct correlation with seminal parameters (count and motility). Alterations of CoQ_{10} content have been shown in conditions associated with male infertility, such as asthenozoospermia and varicocele (Balercia et al. 2009) Coenzyme Q_{10} is the only lipid soluble antioxidant synthesized endogenously. A recent study investigated the role of CoQ_{10} supplementation on semen parameters in idiopathic oligoasthenoteratozoospermia (OAT) (Nadjarzadeh et al. 2011). In the study a total of 47 infertile men with iOAT were randomly assigned to receive 200 mg CoQ_{10} daily or placebo during a 12-week period. The results of the study showed that concentrations of thiobarbituric acid-reactive substances (TBARS) were significantly reduced in serum of treated groups compared with the control. Furthermore, total antioxidant capacity of seminal plasma significantly increased in the CoQ_{10} group. The results provided additional evidence suggesting that CoQ_{10} supplementation is associated with alleviating OS and suggested that CoQ_{10} could be taken as an adjunct therapy in cases of oligoasthenoteratozoospermia. The evaluation of CoQ_{10} supplementation on semen parameters and pregnancy rates in idiopathic oligoasthenoteratozoospermia (iOAT) were also reported in the following year (Safarinejad 2012). These patients were treated with CoQ_{10} 300 mg orally twice daily for 12 months. The results revealed that mean sperm concentration, sperm progressive motility, and sperm with normal morphology improved significantly after a 12-month CoQ_{10} therapy with beneficial effect on pregnancy rate. For the first time, the occurrence of Coenzyme Q_{10} in follicular fluid in females and its relationship with oocyte fertilization and embryo grading was investigated in a study that measured CoQ_{10} levels from 20 infertile women undergoing IVF using HPLC system (Turi et al. 2012). The results of the study showed that CoQ_{10}/Protein levels resulted significantly in mature versus dysmorphic oocytes. Similarly, CoQ_{10}/Cholesterol was significantly higher in grading I–II versus grading III–IV embryos.

3.2.6 L-Carnitine

L-Carnitine (LC) is a small water-soluble molecule which exerts an important function in fat metabolism. L-Carnitine is concentrated in high energy demanding tissues such as skeletal and cardiac muscles and in the epididymis. It plays a significant role in transferring long-chain fatty acids into the mitochondria for oxidation, producing energy (Abdelrazik et al. 2009; Ahmed et al. 2011). The role of L-Carnitine in male infertility has been demonstrated in various studies. It was hypothesised in a study that L-Carnitine helps in maintaining normal fertility (Ahmed et al. 2011). The study was designed to show comparison of seminal free L-Carnitine and sperm quality. Case-controlled convenient sampling was used to assess infertile male subjects from fertile. A total of 61 adult males were selected by consent, and were categorized as fertile and infertile on the basis of history and semen analysis. The results showed that the mean values of sperm count, total motility and normal morphology of asthenozoospermic and oligoasthenoteratozoospermic men were significantly lower when compared with fertile (control). When levels of seminal free L-Carnitine were compared among groups, the result showed that infertile subjects had significantly lower levels when compared to fertile subjects with lowest concentration in azoospermic group, overall suggesting essential role of L-Carnitine level in seminal plasma in maintaining male fertility.

Various studies explored the effect of L-Carnitine in asthenozoospermic patients who underwent IVF-ET. Patients with asthenozoospermia were treated with L-Carnitine 2 g/day. The results demonstrated significantly increased percentage of semen quality (forward motile sperm and sperm concentration) and increased pregnancy rates after the treatment (Wang et al. 2010; Jarow 2003; Lenzi et al. 2003). These studies demonstrated that L-Carnitine supplementation significantly improves two of the most important sperm parameters: concentration and motility for patients undergoing IVF-ET.

The use of L-Carnitine before percutaneous epididymal sperm aspiration (PESA)–ICSI in the treatment of obstructive azoospermia was also explored in a study in which the males were provided with oral 1 g L-Carnitine for 3 months. The result of the experiment revealed that oral medication of L-Carnitine before PESA-ICSI can raise the number and rate of good embryos in obstructive azoospermia patients and therefore benefit the therapeutic outcome (Lu et al. 2010). In another study, L-Carnitine 2 g/day and acetyl-L-Carnitine 1 g/day for 3 months in patients with prostato-vesiculo-epididymitis (PVE) and elevated ROS production, showed that carnitines are an effective treatment in such cases. The results of the study showed increased sperm forward motility and viability in the patients with significant reduction in ROS production (Vicari and Calogero 2001).

Supplementation with L-Carnitine 2 g/day and acetyl-L-Carnitine 1 g/day, appeared to benefit idiopathic infertility in men. According to the study, nearly 22 % of infertile men with enlarged varicose veins in the scrotum who took L-Carnitine and acetyl-L-Carnitine achieved a pregnancy with their partner within 9 months, compared to only 2 % of men taking a placebo (Cavallini et al. 2004).

In females, the protective effect of L-Carnitine was also studied in patients undergoing IVF-ET. The oocytes obtained from the infertile patients were fertilized by ICSI. Embryos were incubated in culture media supplemented with 0.3 mg/mL L-Carnitine and some without L-Carnitine. The results of the study showed significant improvement in the quality of embryos and blastocyst development rate supplemented with L-Carnitine. The authors suggested use of L-Carnitine in culture media could provide a novel approach to improve ICSI outcome in infertile couples (Abdelrazik et al. 2008).

Chapter 4
Role of Combined Antioxidants

Evidence in the literature supported the use of combined antioxidants for the management of free radicals and OS prevention. The effectiveness of combined non-enzymatic antioxidants was investigated in male patients with male factor infertility prior to their third IUI treatment cycle (Dashti et al. 2013). Prior to the treatment, the males were advised to take a combination of tamoxifen 10 mg twice a day, vitamin E 400 IU daily, zinc 15 mg twice a day and selenium 200 mg daily for 3 months. The results of the study detected significant differences in overall values for the four semen parameters (sperm concentration, motility, forward progression and the percentage normal forms) in comparison to the earlier two IUI cycles in the same group. The grouping of the female patients according to their Body Mass Index (BMI) showed crucial differences in pregnancy outcome. The influence of oral antioxidant supplementation on semen quality of IVF patients undergoing IVF/ICSI was analysed in an experiment in which the male patients were supplemented with an oral antioxidant called FertilovitRMplus twice daily for 2 months (Wirleitner et al. 2012). The components of FertilovitRMplus included: Vitamin C 100 mg, Vitamin E 100 mg, Folic acid 500 µg, Zinc 25 mg, Selenium 100 µg, N-acetyl-L-Cysteine 50 mg, L-Carnitine 300 mg, Citrulline 300 mg, Glutathione red 50 mg, Lycopene 4 mg and Coenzyme Q_{10} 15 mg. The results of this study revealed significant reduction in the percentage of immotile sperms in the patients. Additionally, the percentage of class I spermatozoa was significantly higher with drastic improvement in sperm motility as well as in total sperm count.

A study by Tunc's group added to the already growing body of evidence supporting the use of antioxidant combinational therapy to improve sperm DNA integrity, especially for men undergoing IVF-ICSI treatment (Tunc et al. 2009). In their study, a total of 50 infertile men identified with OS were administered with an oral antioxidant therapy called Menevit for a period of 3 months. The components included Lycopene 6 mg, Vitamin E 400 IU, Vitamin C 100 mg, Zinc 25 mg, Selenium 26 g, Folate 500 g, and Garlic oil 333 g (equivalent 1 g garlic). The results of the study suggested that treatment of men with a high degree of oxidative DNA damage with antioxidants before their partner commences IVF-ICSI therapy, may be capable of improving pregnancy outcomes.

In a previous study, the role of Menevit antioxidant therapy was examined on embryo quality and pregnancy outcome during IVF/ICSI treatment (Tremellen et al. 2007). Male participants were randomly assigned to take either one capsule per day of the Menevit antioxidant or an identical-in-appearance placebo for 3 months prior to their partner's IVF cycle. The results of the study demonstrated a statistically significant improvement in viable pregnancy rate compared to the control group, In another study, zinc therapy in combination with vitamin E or with vitamins E+C were associated with comparably improved sperm parameters with less OS, sperm apoptosis and sperm DNA fragmentation index (DFI) (Omu et al. 2008). Men with asthenozoospermia were orally provided with Zinc 5 mg, Vitamin E+Zinc 10 mg and Zinc+Vitamins E+C 200 mg for 3 months. The results of the study demonstrated that zinc therapy alone, in combination with Vitamin E or with Vitamins E+C were associated with comparably improved sperm parameters. Overall, it was concluded that this combined therapy reduced asthenozoospermia through several mechanisms, such as prevention of OS, apoptosis and sperm DNA fragmentation.

Rizzo's group evaluated the efficacy of a treatment with myo-inositol plus folic acid plus melatonin compared with myo-inositol plus folic acid alone on oocyte quality in women who underwent IVF (Rizzo et al. 2010). Patients were assigned to obtain either 2 g myo-inositol twice a day combined with 200 mg folic acid and 3 mg melatonin. The data showed that although the number of oocytes retrieved did not differ between the two treatment groups, but women co-treated with melatonin had an improvement in oocyte quality and a higher mean number of morphologically mature oocytes at ovum pick up. Women co-treated with melatonin were also found to have a greater number of top-quality embryos compared to those treated only with myo-inositol plus folic acid.

Chapter 5
Conclusion

Reactive oxygen species (ROS) are present in the reproductive organs and its fluids in the female, and is generated by immature sperm, leukocytes from seminal plasma and by the presence of varicocele in the male. While ROS play an important physiological role in the process of reproduction, however, excessive amounts of ROS leads to a state of oxidative stress, lipid peroxidation and subsequently DNA damage. Spermatozoa are especially vulnerable to the effects of ROS as it lacks antioxidant defences. Oral antioxidant supplementation could potentially help quench the increased levels of oxidative stress and improve the quality of gametes produced, especially in the male. Antioxidant therapies used clinically as oral supplementation include vitamin E, vitamin C, selenium, folic acid and Coenzyme Q_{10} either alone or in combination.

Couples suffering from reproductive disorders and/or infertility who seek assisted reproduction are often plighted by the harmful effects of oxidative stress on gametes and the developing embryo, and its subsequent negative impact on the pregnancy outcome. During assisted reproduction, oxidative stress could be generated endogenously by the gametes itself or exogenously during sperm or oocyte preparation, gamete handling and transfer of the embryo. Strategies to minimize oxidative stress during *in vitro* fertilization and embryo transfer include the addition of antioxidants to the culture/handling media and mindful practices during the mechanical and laboratory techniques involved, as illustrated in Fig. 5.1. Although various studies using enzymatic and non-enzymatic antioxidants have reported improvement in sperm parameters and ART outcome, but to date, no one antioxidant in particular or any certain combination of antioxidants could be proposed as the antidote to oxidative stress for infertile couples seeking assisted reproduction. There is a need for an increased number of well-designed randomized controlled trials in different populations to determine which of the antioxidant treatments would present with the best outcome in subfertile couples seeking assisted reproduction.

© The Author 2014
A. Agarwal et al., *Strategies to Ameliorate Oxidative Stress During
Assisted Reproduction*, SpringerBriefs in Reproductive Biology,
DOI 10.1007/978-3-319-10259-7_5

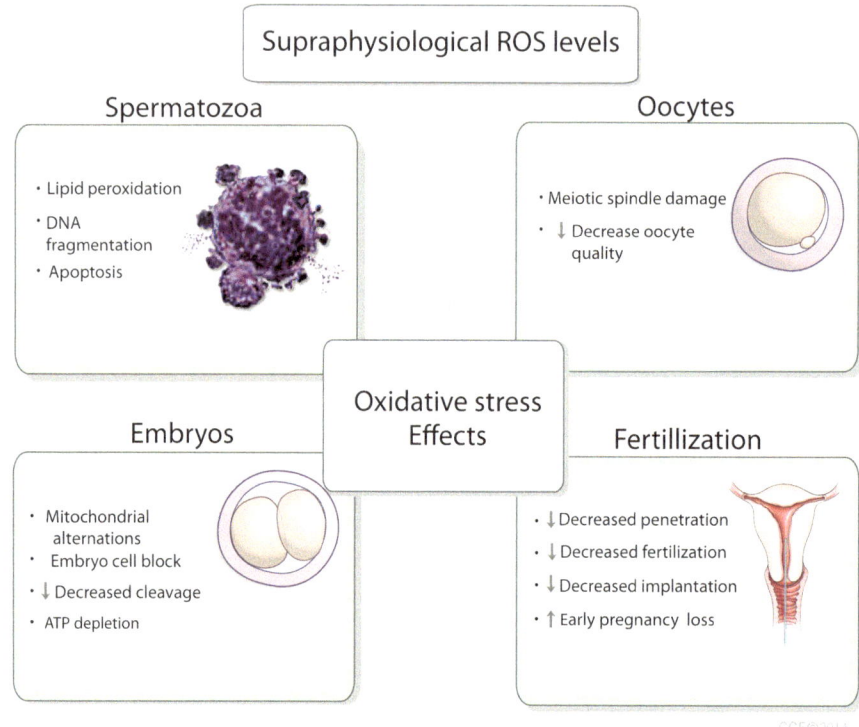

Fig. 5.1 Sources of oxidative stress and interventions to overcome its effects during assisted reproductive technology

ROS reactive oxygen species, *ATP* adenosine triphosphate, *DNA* deoxyribonucleic acid

References

Abdelrazik H, El-Damen A, Badrawy H, Sharma R, Agarwal A. L-Carnitine improves embryo quality and increases blastocyst development rate in couples undergoing ICSI. In: 64th annual meeting of the American Society of Reproductive Medicine, 8–12 Nov 2008, San Francisco, CA.

Abdelrazik H, Sharma R, Mahfouz R, Agarwal A. L-carnitine decreases DNA damage and improves the in vitro blastocyst development rate in mouse embryos. Fertil Steril. 2009;91:589–96.

Agarwal A, Allamaneni SS. Role of free radicals in female reproductive diseases and assisted reproduction. Reprod Biomed Online. 2004;9:338–47.

Agarwal A, Prabakaran SA. Mechanism, measurement, and prevention of oxidative stress in male reproductive physiology. Indian J Exp Biol. 2005;43:963–74.

Agarwal A, Ikemoto I, Loughlin KR. Effect of sperm washing on levels of reactive oxygen species in semen. Arch Androl. 1994;33:157–62.

Agarwal A, Saleh RA, Bedaiwy MA. Role of reactive oxygen species in the pathophysiology of human reproduction. Fertil Steril. 2003;79:829–43.

Agarwal A, Nallella KP, Allamaneni SS, Said TM. Role of antioxidants in treatment of male infertility: an overview of the literature. Reprod Biomed Online. 2004;8:616–27.

Agarwal A, Gupta S, Sharma R. Oxidative stress and its implications in female infertility – a clinician's perspective. Reprod Biomed Online. 2005a;11:641–50.

Agarwal A, Gupta S, Sharma RK. Role of oxidative stress in female reproduction. Reprod Biol Endocrinol. 2005b;3:28.

Agarwal A, Gupta S, Sikka S. The role of free radicals and antioxidants in reproduction. Curr Opin Obstet Gynecol. 2006a;18:325–32.

Agarwal A, Said TM, Bedaiwy MA, Banerjee J, Alvarez JG. Oxidative stress in an assisted reproductive techniques setting. Fertil Steril. 2006b;86:503–12.

Agarwal A, Sharma RK, Nallella KP, Jr. Thomas AJ, Alvarez JG, Sikka SC. Reactive oxygen species as an independent marker of male factor infertility. Fertil Steril. 2006c;86:878–85.

Agarwal A, Aponte-Mellado A, Premkumar BJ, Shaman A, Gupta S. The effects of oxidative stress on female reproduction: a review. Reprod Biol Endocrinol. 2012;10:49.

Ahmed SD, Karira KA, Jagdesh, Ahsan S. Role of L-carnitine in male infertility. J Pak Med Assoc. 2011;61:732–6.

Aitken RJ, Krausz C. Oxidative stress, DNA damage and the Y chromosome. Reproduction. 2001;122:497–506.

Akmal M, Qadri JQ, Al-Waili NS, Thangal S, Haq A, Saloom KY. Improvement in human semen quality after oral supplementation of vitamin C. J Med Food. 2006;9:440–2.

© The Author 2014
A. Agarwal et al., *Strategies to Ameliorate Oxidative Stress During Assisted Reproduction*, SpringerBriefs in Reproductive Biology, DOI 10.1007/978-3-319-10259-7

Alscher RG, Erturk N, Heath LS. Role of superoxide dismutases (SODs) in controlling oxidative stress in plants. J Exp Bot. 2002;53:1331–41.

Alvarez JG, Storey BT. Taurine, hypotaurine, epinephrine and albumin inhibit lipid peroxidation in rabbit spermatozoa and protect against loss of motility. Biol Reprod. 1983;29:548–55.

Alvarez JG, Touchstone JC, Blasco L, Storey BT. Spontaneous lipid peroxidation and production of hydrogen peroxide and superoxide in human spermatozoa. Superoxide dismutase as major enzyme protectant against oxygen toxicity. J Androl. 1987;8:338–48.

Alvarez JG, Lasso JL, Blasco L, Nunez RC, Heyner S, Caballero PP, Storey BT. Centrifugation of human spermatozoa induces sublethal damage; separation of human spermatozoa from seminal plasma by a dextran swim-up procedure without centrifugation extends their motile lifetime. Hum Reprod. 1993;8:1087–92.

Ambekar AS, Nirujogi RS, Srikanth SM, Chavan S, Kelkar DS, Hinduja I, Zaveri K, Prasad TS, Harsha HC, Pandey A, Mukherjee S. Proteomic analysis of human follicular fluid: a new perspective towards understanding folliculogenesis. J Proteomics. 2013;87:68–77.

Anger JT, Gilbert BR, Goldstein M. Cryopreservation of sperm: indications, methods and results. J Urol. 2003;170:1079–84.

Askoxylaki M, Siristatidis C, Chrelias C, Vogiatzi P, Creatsa M, Salamalekis G, Vrantza T, Vrachnis N, Kassanos D. Reactive oxygen species in the follicular fluid of subfertile women undergoing in vitro fertilization: a short narrative review. J Endocrinol Invest. 2013;36:1117–20.

Atig F, Raffa M, Ali HB, Abdelhamid K, Saad A, Ajina M. Altered antioxidant status and increased lipid per-oxidation in seminal plasma of Tunisian infertile men. Int J Biol Sci. 2012;8: 139–49.

Attaran M, Pasqualotto E, Falcone T, Goldberg JM, Miller KF, Agarwal A, Sharma RK. The effect of follicular fluid reactive oxygen species on the outcome of in vitro fertilization. Int J Fertil Womens Med. 2000;45:314–20.

Augustine LM, Markelewicz Jr RJ, Boekelheide K, Cherrington NJ. Xenobiotic and endobiotic transporter mRNA expression in the blood-testis barrier. Drug Metab Dispos. 2005;33:182–9.

Bagger PV, Byskov AG, Christiansen MD. Maturation of mouse oocytes in vitro is influenced by alkalization during their isolation. J Reprod Fertil. 1987;80:251–5.

Baker MA, Aitken RJ. Reactive oxygen species in spermatozoa: methods for monitoring and significance for the origins of genetic disease and infertility. Reprod Biol Endocrinol. 2005;3:67.

Balercia G, Mancini A, Paggi F, Tiano L, Pontecorvi A, Boscaro M, Lenzi A, Littarru GP. Coenzyme Q10 and male infertility. J Endocrinol Invest. 2009;32:626–32.

Barrionuevo MJ, Schwandt RA, Rao PS, Graham LB, Maisel LP, Yeko TR. Nitric oxide (NO) and interleukin-1beta (IL-1beta) in follicular fluid and their correlation with fertilization and embryo cleavage. Am J Reprod Immunol. 2000;44:359–64.

Batioglu AS, Sahin U, Gurlek B, Ozturk N, Unsal E. The efficacy of melatonin administration on oocyte quality. Gynecol Endocrinol. 2012;28:91–3.

Bedaiwy MA, Falcone T, Mohamed MS, Aleem AA, Sharma RK, Worley SE, Thornton J, Agarwal A. Differential growth of human embryos in vitro: role of reactive oxygen species. Fertil Steril. 2004;82:593–600.

Bedaiwy MA, Mahfouz RZ, Goldberg JM, Sharma R, Falcone T, Abdel Hafez MF, Agarwal A. Relationship of reactive oxygen species levels in day 3 culture media to the outcome of in vitro fertilization/intracytoplasmic sperm injection cycles. Fertil Steril. 2010;94:2037–42.

Bedaiwy MA, Elnashar SA, Goldberg JM, Sharma R, Mascha EJ, Arrigain S, Agarwal A, Falcone T. Effect of follicular fluid oxidative stress parameters on intracytoplasmic sperm injection outcome. Gynecol Endocrinol. 2012;28:51–5.

Bedford JM, Dobrenis A. Light exposure of oocytes and pregnancy rates after their transfer in the rabbit. J Reprod Fertil. 1989;85:477–81.

Behrman HR, Kodaman PH, Preston SL, Gao S. Oxidative stress and the ovary. J Soc Gynecol Investig. 2001;8:S40–2.

Benchaib M, Braun V, Lornage J, Hadj S, Salle B, Lejeune H, Guerin JF. Sperm DNA fragmentation decreases the pregnancy rate in an assisted reproductive technique. Hum Reprod. 2003;18:1023–8.

Bontekoe S, Mantikou E, Van Wely M, Seshadri S, Repping S, Mastenbroek S. Low oxygen concentrations for embryo culture in assisted reproductive technologies. Cochrane Database Syst Rev. 2012;7:CD008950.

Borowiecka M, Wojsiat J, Polac I, Radwan M, Radwan P, Zbikowska HM. Oxidative stress markers in follicular fluid of women undergoing in vitro fertilization and embryo transfer. Syst Biol Reprod Med. 2012;58:301–5.

Bozhedomov VA, Gromenko DS, Ushakova IV, Toroptseva MV, Galimov ShN, Alekcandrova LA, Teodorovich OV, Sukhikh TG. [Oxidative stress of spermatozoa in pathogenesis of male infertility]. Urologiia. 2009;(2):51–6.

Branco CS, Garcez ME, Pasqualotto FF, Erdtman B, Salvador M. Resveratrol and ascorbic acid prevent DNA damage induced by cryopreservation in human semen. Cryobiology. 2010;60:235–7.

Buser P, Imbert M, Buser PA. Vision. Cambridge, MA: Bradford, MIT Press; 1992.

Calzi F, Papaleo E, Rabellotti E, Ottolina J, Vailati S, Viganò P, Candiani M. Exposure of embryos to oxygen at low concentration in a cleavage stage transfer program: reproductive outcomes in a time-series analysis. Clin Lab. 2012;58(9–10):997–1003.

Catt JW, Henman M. Toxic effects of oxygen on human embryo development. Hum Reprod. 2000;15 Suppl 2:199–206.

Cavallini G, Ferraretti AP, Gianaroli L, Biagiotti G, Vitali G. Cinnoxicam and L-carnitine/acetyl-L-carnitine treatment for idiopathic and varicocele-associated oligoasthenospermia. J Androl. 2004;25:761–70. Discussion 771–2.

Chattopadhayay R, Ganesh A, Samanta J, Jana SK, Chakravarty BN, Chaudhury K. Effect of follicular fluid oxidative stress on meiotic spindle formation in infertile women with polycystic ovarian syndrome. Gynecol Obstet Invest. 2010;69:197–202.

Chen C, Han S, Liu W, Wang Y, Huang G. Effect of vitrification on mitochondrial membrane potential in human metaphase II oocytes. J Assist Reprod Genet. 2012;29:1045–50.

Chi HJ, Kim JH, Ryu CS, Lee JY, Park JS, Chung DY, Choi SY, Kim MH, Chun EK, Roh SI. Protective effect of antioxidant supplementation in sperm-preparation medium against oxidative stress in human spermatozoa. Hum Reprod. 2008;23:1023–8.

Cicek N, Eryilmaz OG, Sarikaya E, Gulerman C, Genc Y. Vitamin E effect on controlled ovarian stimulation of unexplained infertile women. J Assist Reprod Genet. 2012;29:325–8.

Ciotta L, Stracquadanio M, Pagano I, Carbonaro A, Palumbo M, Gulino F. Effects of myo-inositol supplementation on oocyte's quality in PCOS patients: a double blind trial. Eur Rev Med Pharmacol Sci. 2011;15:509–14.

Cohen J, Gilligan A, Esposito W, Schimmel T, Dale B. Ambient air and its potential effects on conception in vitro. Hum Reprod. 1997;12:1742–9.

Colagar AH, Marzony ET. Ascorbic acid in human seminal plasma: determination and its relationship to sperm quality. J Clin Biochem Nutr. 2009;45:144–9.

Comhaire FH, Christophe AB, Zalata AA, Dhooge WS, Mahmoud AM, Depuydt CE. The effects of combined conventional treatment, oral antioxidants and essential fatty acids on sperm biology in subfertile men. Prostaglandins Leukot Essent Fatty Acids. 2000;63:159–65.

Crha I, Hruba D, Ventruba P, Fiala J, Totusek J, Visnova H. Ascorbic acid and infertility treatment. Cent Eur J Public Health. 2003;11:63–7.

Crisol L, Matorras R, Aspichueta F, Exposito A, Hernandez ML, Ruiz-Larrea MB, Mendoza R, Ruiz-Sanz JI. Glutathione peroxidase activity in seminal plasma and its relationship to classical sperm parameters and in vitro fertilization-intracytoplasmic sperm injection outcome. Fertil Steril. 2012;97:852–7.

Dalzell LH, Mcvicar CM, Mcclure N, Lutton D, Lewis SE. Effects of short and long incubations on DNA fragmentation of testicular sperm. Fertil Steril. 2004;82:1443–5.

Das S, Chattopadhyay R, Ghosh S, Goswami SK, Chakravarty BN, Chaudhury K. Reactive oxygen species level in follicular fluid – embryo quality marker in IVF? Hum Reprod. 2006;21:2403–7.

Dashti MAG, Alhamar AY, Shawky H, Bakhiet M. Effectiveness of combined empirical therapies and double IUI procedures in treatment of male factor infertility. Andrology. 2013;2:112. doi:10.4172/2167-0250.1000112.

Davidson JF, Schiestl RH. Mitochondrial respiratory electron carriers are involved in oxidative stress during heat stress in Saccharomyces cerevisiae. Mol Cell Biol. 2001;21:8483–9.

de Lamirande E, Gagnon C. A positive role for the superoxide anion in triggering hyperactivation and capacitation of human spermatozoa. Int J Androl. 1993;16:21–5.

de Lamirande E, Gagnon C. Capacitation-associated production of superoxide anion by human spermatozoa. Free Radic Biol Med. 1995;18:487–95.

Di Santo M, Tarozzi N, Nadalini M, Borini A. Human sperm cryopreservation: update on techniques, effect on DNA integrity, and implications for ART. Adv Urol. 2012;2012:854837.

Doshi SB, Khullar K, Sharma RK, Agarwal A. Role of reactive nitrogen species in male infertility. Reprod Biol Endocrinol. 2012;10:109.

Du Plessis S, Makker K, Desai NR, Agarwal A. Impact of oxidative stress on IVF. Obstet Gynecol. 2008;34:539–54.

Ebisch IM, Thomas CM, Peters WH, Braat DD, Steegers-Theunissen RP. The importance of folate, zinc and antioxidants in the pathogenesis and prevention of subfertility. Hum Reprod Update. 2007;13:163–74.

Edwards AM, Silva E. Effect of visible light on selected enzymes, vitamins and amino acids. J Photochem Photobiol B. 2001;63:126–31.

Eichler M, Lavi R, Shainberg A, Lubart R. Flavins are source of visible-light-induced free radical formation in cells. Lasers Surg Med. 2005;37:314–9.

Eryilmaz OG, Devran A, Sarikaya E, Aksakal FN, Mollamahmutoglu L, Cicek N. Melatonin improves the oocyte and the embryo in IVF patients with sleep disturbances, but does not improve the sleeping problems. J Assist Reprod Genet. 2011;28:815–20.

Ferguson WJ, Braunschweiger KI, Braunschweiger WR, Smith JR, Mccormick JJ, Wasmann CC, Jarvis NP, Bell DH, Good NE. Hydrogen ion buffers for biological research. Anal Biochem. 1980;104:300–10.

Ford WC, Whittington K, Williams AC. Reactive oxygen species in human sperm suspensions: production by leukocytes and the generation of NADPH to protect sperm against their effects. Int J Androl. 1997;20 Suppl 3:44–9.

Gandini L, Lombardo F, Paoli D, Caponecchia L, Familiari G, Verlengia C, Dondero F, Lenzi A. Study of apoptotic DNA fragmentation in human spermatozoa. Hum Reprod. 2000;15:830–9.

Garrido N, Meseguer M, Alvarez J, Simon C, Pellicer A, Remohi J. Relationship among standard semen parameters, glutathione peroxidase/glutathione reductase activity, and mRNA expression and reduced glutathione content in ejaculated spermatozoa from fertile and infertile men. Fertil Steril. 2004;82 Suppl 3:1059–66.

Gavella M, Lipovac V. NADH-dependent oxidoreductase (diaphorase) activity and isozyme pattern of sperm in infertile men. Arch Androl. 1992;28:135–41.

Geva E, Bartoov B, Zabludovsky N, Lessing JB, Lerner-Geva L, Amit A. The effect of antioxidant treatment on human spermatozoa and fertilization rate in an in vitro fertilization program. Fertil Steril. 1996;66:430–4.

Gharagozloo P, Aitken RJ. The role of sperm oxidative stress in male infertility and the significance of oral antioxidant therapy. Hum Reprod. 2011;26:1628–40.

Giraud MN, Motta C, Boucher D, Grizard G. Membrane fluidity predicts the outcome of cryopreservation of human spermatozoa. Hum Reprod. 2000;15:2160–4.

Girotti AW. Photosensitized oxidation of membrane lipids: reaction pathways, cytotoxic effects, and cytoprotective mechanisms. J Photochem Photobiol B. 2001;63:103–13.

Gomes-Sobrinho DB, Oliveira JB, Petersen CG, Mauri AL, Silva LF, Massaro FC, Baruffi RL, Cavagna M, Franco Jr JG. IVF/ICSI outcomes after culture of human embryos at low oxygen tension: a meta-analysis. Reprod Biol Endocrinol. 2011;9:143.

Greco E, Iacobelli M, Rienzi L, Ubaldi F, Ferrero S, Tesarik J. Reduction of the incidence of sperm DNA fragmentation by oral antioxidant treatment. J Androl. 2005a;26:349–53.

Greco E, Romano S, Iacobelli M, Ferrero S, Baroni E, Minasi MG, Ubaldi F, Rienzi L, Tesarik J. ICSI in cases of sperm DNA damage: beneficial effect of oral antioxidant treatment. Hum Reprod. 2005b;20:2590–4.

Gualtieri R, Iaccarino M, Mollo V, Prisco M, Iaccarino S, Talevi R. Slow cooling of human oocytes: ultrastructural injuries and apoptotic status. Fertil Steril. 2009;91:1023–34.

Guerin P, El Mouatassim S, Menezo Y. Oxidative stress and protection against reactive oxygen species in the pre-implantation embryo and its surroundings. Hum Reprod Update. 2001;7: 175–89.

Halliwell B, Gutteridge JM. Free radicals and antioxidant protection: mechanisms and significance in toxicology and disease. Hum Toxicol. 1988;7:7–13.

Hammadeh ME, Al Hasani S, Rosenbaum P, Schmidt W, Fischer Hammadeh C. Reactive oxygen species, total antioxidant concentration of seminal plasma and their effect on sperm parameters and outcome of IVF/ICSI patients. Arch Gynecol Obstet. 2008;277:515–26.

Henkel RR, Schill WB. Sperm preparation for ART. Reprod Biol Endocrinol. 2003;1:108.

Henmi H, Endo T, Kitajima Y, Manase K, Hata H, Kudo R. Effects of ascorbic acid supplementation on serum progesterone levels in patients with a luteal phase defect. Fertil Steril. 2003;80:459–61.

Hockberger PE, Skimina TA, Centonze VE, Lavin C, Chu S, Dadras S, Reddy JK, White JG. Activation of flavin-containing oxidases underlies light-induced production of H2O2 in mammalian cells. Proc Natl Acad Sci U S A. 1999;96:6255–60.

Host E, Lindenberg S, Smidt-Jensen S. The role of DNA strand breaks in human spermatozoa used for IVF and ICSI. Acta Obstet Gynecol Scand. 2000;79:559–63.

Hu Y, Betzendahl I, Cortvrindt R, Smitz J, Eichenlaub-Ritter U. Effects of low O2 and ageing on spindles and chromosomes in mouse oocytes from pre-antral follicle culture. Hum Reprod. 2001;16:737–48.

Huntriss J, Picton HM. Epigenetic consequences of assisted reproduction and infertility on the human preimplantation embryo. Hum Fertil (Camb). 2008;11:85–94.

Irvine DS, Twigg JP, Gordon EL, Fulton N, Milne PA, Aitken RJ. DNA integrity in human spermatozoa: relationships with semen quality. J Androl. 2000;21:33–44.

Jain SK, Mcvie R, Smith T. Vitamin E supplementation restores glutathione and malondialdehyde to normal concentrations in erythrocytes of type 1 diabetic children. Diabetes Care. 2000;23:1389–94.

Jana SK, K NB, Chattopadhyay R, Chakravarty B, Chaudhury K. Upper control limit of reactive oxygen species in follicular fluid beyond which viable embryo formation is not favorable. Reprod Toxicol. 2010;29:447–51.

Jarow JP. Use of carnitine therapy in selected cases of male factor infertility: a double-blind cross-over trial. J Urol. 2003;170:677.

Jenkins TG, Aston KI, Carrell DT. Supplementation of cryomedium with ascorbic acid-2-glucoside (AA2G) improves human sperm post-thaw motility. Fertil Steril. 2011;95:2001–4.

Jones DP. The role of oxygen concentration in oxidative stress: hypoxic and hyperoxic models. In: Sies H, editor. Oxidative stress. New York: Academic; 1985.

Jones A, Van Blerkom J, Davis P, Toledo AA. Cryopreservation of metaphase II human oocytes effects mitochondrial membrane potential: implications for developmental competence. Hum Reprod. 2004;19:1861–6.

Joshi R, Adhikari S, Patro BS, Chattopadhyay S, Mukherjee T. Free radical scavenging behavior of folic acid: evidence for possible antioxidant activity. Free Radic Biol Med. 2001;30:1390–9.

Kalthur G, Raj S, Thiyagarajan A, Kumar S, Kumar P, Adiga SK. Vitamin E supplementation in semen-freezing medium improves the motility and protects sperm from freeze-thaw-induced DNA damage. Fertil Steril. 2011;95:1149–51.

Karuputhula NB, Chattopadhyay R, Chakravarty B, Chaudhury K. Oxidative status in granulosa cells of infertile women undergoing IVF. Syst Biol Reprod Med. 2013;59:91–8.

Keskes-Ammar L, Feki-Chakroun N, Rebai T, Sahnoun Z, Ghozzi H, Hammami S, Zghal K, Fki H, Damak J, Bahloul A. Sperm oxidative stress and the effect of an oral vitamin E and selenium supplement on semen quality in infertile men. Arch Androl. 2003;49:83–94.

Khademi A, Alleyassin A, Safdarian L, Hamed EA, Rabiee E, Haghaninezhad H. The effects of L-carnitine on sperm parameters in smoker and non-smoker patients with idiopathic sperm abnormalities. J Assist Reprod Genet. 2005;22:395–9.

Kovacic B, Vlaisavljevic V. Influence of atmospheric versus reduced oxygen concentration on development of human blastocysts in vitro: a prospective study on sibling oocytes. Reprod Biomed Online. 2008;17:229–36.

Kovacic B, Sajko MC, Vlaisavljevic V. A prospective, randomized trial on the effect of atmospheric versus reduced oxygen concentration on the outcome of intracytoplasmic sperm injection cycles. Fertil Steril. 2010;94:511–9.

Lampiao F. Free radicals generation in an in vitro fertilization setting and how to minimize them. World J Obstet Gynecol. 2012;1:29–34.

Lampiao F, Strijdom H, Du Plessis SS. Effects of sperm processing techniques involving centrifugation on nitric oxide, reactive oxygen species generation and sperm function. Open Androl J. 2010;2:1–5.

Larkindale J, Knight MR. Protection against heat stress-induced oxidative damage in Arabidopsis involves calcium, abscisic acid, ethylene, and salicylic acid. Plant Physiol. 2002;128:682–95.

Lee TH, Lee MS, Liu CH, Tsao HM, Huang CC, Yang YS. The association between microenvironmental reactive oxygen species and embryo development in assisted reproduction technology cycles. Reprod Sci. 2012;19:725–32.

Lenzi A, Lombardo F, Sgro P, Salacone P, Caponecchia L, Dondero F, Gandini L. Use of carnitine therapy in selected cases of male factor infertility: a double-blind crossover trial. Fertil Steril. 2003;79:292–300.

Li Z, Lin Q, Liu R, Xiao W, Liu W. Protective effects of ascorbate and catalase on human spermatozoa during cryopreservation. J Androl. 2010;31:437–44.

Liu J, Li Y. [Effect of oxidative stress and apoptosis in granulosa cells on the outcome of IVF-ET]. Zhong Nan Da Xue Xue Bao Yi Xue Ban. 2010;35:990–4.

Liu F, He L, Liu Y, Shi Y, Du H. The expression and role of oxidative stress markers in the serum and follicular fluid of patients with endometriosis. Clin Exp Obstet Gynecol. 2013;40:372–6.

Lu SM, Li X, Zhang HB, Hu JM, Yan JH, Liu JL, Chen ZJ. [Use of L-carnitine before percutaneous epididymal sperm aspiration-intracytoplasmic sperm injection for obstructive azoospermia]. Zhonghua Nan Ke Xue. 2010;16:919–21.

Marquez B, Suarez SS. Bovine sperm hyperactivation is promoted by alkaline-stimulated Ca2+ influx. Biol Reprod. 2007;76:660–5.

Martin-Romero FJ, Miguel-Lasobras EM, Dominguez-Arroyo JA, Gonzalez-Carrera E, Alvarez IS. Contribution of culture media to oxidative stress and its effect on human oocytes. Reprod Biomed Online. 2008;17:652–61.

Marzec-Wroblewska U, Kaminski P, Lakota P, Szymanski M, Wasilow K, Ludwikowski G, Kuligowska-Prusinska M, Odrowaz-Sypniewska G, Stuczynski T, Michalkiewicz J. Zinc and iron concentration and SOD activity in human semen and seminal plasma. Biol Trace Elem Res. 2011;143:167–77.

Matos L, Stevenson D, Gomes F, Silva-Carvalho JL, Almeida H. Superoxide dismutase expression in human cumulus oophorus cells. Mol Hum Reprod. 2009;15:411–9.

Mckinney KA, Lewis SE, Thompson W. The effects of pentoxifylline on the generation of reactive oxygen species and lipid peroxidation in human spermatozoa. Andrologia. 1996;28:15–20.

Mehta A, Sigman M. Identification and preparation of sperm for ART. Urol Clin North Am. 2014;41:169–80.

Menezo YJ, Hazout A, Panteix G, Robert F, Rollet J, Cohen-Bacrie P, Chapuis F, Clement P, Benkhalifa M. Antioxidants to reduce sperm DNA fragmentation: an unexpected adverse effect. Reprod Biomed Online. 2007;14:418–21.

Mora-Esteves C, Shin D. Nutrient supplementation: improving male fertility fourfold. Semin Reprod Med. 2013;31:293–300.

Moshkdanian G, Nematollahi-Mahani SN, Pouya F, Nematollahi-Mahani A. Antioxidants rescue stressed embryos at a rate comparable with co-culturing of embryos with human umbilical cord mesenchymal cells. J Assist Reprod Genet. 2011;28:343–9.

Moslemi MK, Tavanbakhsh S. Selenium-vitamin E supplementation in infertile men: effects on semen parameters and pregnancy rate. Int J Gen Med. 2011;4:99–104.

Mostafa T, Tawadrous G, Roaia MM, Amer MK, Kader RA, Aziz A. Effect of smoking on seminal plasma ascorbic acid in infertile and fertile males. Andrologia. 2006;38:221–4.

Moubasher AE, El Din AM, Ali ME, El-Sherif WT, Gaber HD. Catalase improves motility, vitality and DNA integrity of cryopreserved human spermatozoa. Andrologia. 2013;45:135–9.

Nadjarzadeh A, Sadeghi MR, Amirjannati N, Vafa MR, Motevalian SA, Gohari MR, Akhondi MA, Yavari P, Shidfar F. Coenzyme Q10 improves seminal oxidative defense but does not affect on semen parameters in idiopathic oligoasthenoteratozoospermia: a randomized double-blind, placebo controlled trial. J Endocrinol Invest. 2011;34:e224–8.

Nakamura Y, Tamura H, Takayama H, Kato H. Increased endogenous level of melatonin in preovulatory human follicles does not directly influence progesterone production. Fertil Steril. 2003;80:1012–6.

Nasr-Esfahani MH, Johnson MH. How does transferrin overcome the in vitro block to development of the mouse preimplantation embryo? J Reprod Fertil. 1992;96:41–8.

O'Bryan MK, Zini A, Cheng CY, Schlegel PN. Human sperm endothelial nitric oxide synthase expression: correlation with sperm motility. Fertil Steril. 1998;70:1143–7.

Omu AE, Al-Azemi MK, Kehinde EO, Anim JT, Oriowo MA, Mathew TC. Indications of the mechanisms involved in improved sperm parameters by zinc therapy. Med Princ Pract. 2008;17:108–16.

Oral O, Kutlu T, Aksoy E, Ficicioglu C, Uslu H, Tugrul S. The effects of oxidative stress on outcomes of assisted reproductive techniques. J Assist Reprod Genet. 2006;23:81–5.

Orsi NM, Leese HJ. Protection against reactive oxygen species during mouse preimplantation embryo development: role of EDTA, oxygen tension, catalase, superoxide dismutase and pyruvate. Mol Reprod Dev. 2001;59:44–53.

Ortiz A, Espino J, Bejarano I, Lozano GM, Monllor F, Garcia JF, Pariente JA, Rodriguez AB. High endogenous melatonin concentrations enhance sperm quality and short-term in vitro exposure to melatonin improves aspects of sperm motility. J Pineal Res. 2011;50:132–9.

Otsuki J, Nagai Y, Matsuyama Y, Terada T, Era S. The influence of the redox state of follicular fluid albumin on the viability of aspirated human oocytes. Syst Biol Reprod Med. 2012;58:149–53.

Ottosen LD, Hindkjaer J, Ingerslev J. Light exposure of the ovum and preimplantation embryo during ART procedures. J Assist Reprod Genet. 2007;24:99–103.

Oyawoye O, Abdel Gadir A, Garner A, Constantinovici N, Perrett C, Hardiman P. Antioxidants and reactive oxygen species in follicular fluid of women undergoing IVF: relationship to outcome. Hum Reprod. 2003;18:2270–4.

Ozawa M, Nagai T, Fahrudin M, Karja NW, Kaneko H, Noguchi J, Ohnuma K, Kikuchi K. Addition of glutathione or thioredoxin to culture medium reduces intracellular redox status of porcine IVM/IVF embryos, resulting in improved development to the blastocyst stage. Mol Reprod Dev. 2006;73:998–1007.

Papaleo E, Unfer V, Baillargeon JP, Chiu TT. Contribution of myo-inositol to reproduction. Eur J Obstet Gynecol Reprod Biol. 2009;147:120–3.

Papaleo E, Potenza MT, Brigante C, De Michele E, Pellegrino M, Candiani M. Nutrients and infertility: an alternative perspective. Eur Rev Med Pharmacol Sci. 2011;15:515–7.

Pasqualotto EB, Agarwal A, Sharma RK, Izzo VM, Pinotti JA, Joshi NJ, Rose BI. Effect of oxidative stress in follicular fluid on the outcome of assisted reproductive procedures. Fertil Steril. 2004;81:973–6.

Pasqualotto EB, Lara LV, Salvador M, Sobreiro BP, Borges E, Pasqualotto FF. The role of enzymatic antioxidants detected in the follicular fluid and semen of infertile couples undergoing assisted reproduction. Hum Fertil (Camb). 2009;12:166–71.

Phillips KP, Leveille MC, Claman P, Baltz JM. Intracellular pH regulation in human preimplantation embryos. Hum Reprod. 2000;15:896–904.

Plante M, de Lamirande E, Gagnon C. Reactive oxygen species released by activated neutrophils, but not by deficient spermatozoa, are sufficient to affect normal sperm motility. Fertil Steril. 1994;62:387–93.

Polak G, Koziol-Montewka M, Gogacz M, Blaszkowska I, Kotarski J. Total antioxidant status of peritoneal fluid in infertile women. Eur J Obstet Gynecol Reprod Biol. 2001a;94:261–3.

Polak G, Koziol-Montewka M, Tarkowski R, Kotarski J. [Peritoneal fluid and plasma 4-hydroxynonenal and malonyldialdehyde concentrations in infertile women]. Ginekol Pol. 2001b;72:1316–20.

Rahimi G, Isachenko E, Sauer H, Isachenko V, Wartenberg M, Hescheler J, Mallmann P, Nawroth F. Effect of different vitrification protocols for human ovarian tissue on reactive oxygen species and apoptosis. Reprod Fertil Dev. 2003;15:343–9.

Rajani S, Chattopadhyay R, Goswami SK, Ghosh S, Sharma S, Chakravarty B. Assessment of oocyte quality in polycystic ovarian syndrome and endometriosis by spindle imaging and reactive oxygen species levels in follicular fluid and its relationship with IVF-ET outcome. J Hum Reprod Sci. 2012;5:187–93.

Rakhit M, Gokul SR, Agarwal A, du Plessis SS. Antioxidant strategies to overcome OS in IVF-Embryo transfer. In: Studies on Women's Health, 237–262. Humana Press; 2013.

Revelli A, Delle Piane L, Casano S, Molinari E, Massobrio M, Rinaudo P. Follicular fluid content and oocyte quality: from single biochemical markers to metabolomics. Reprod Biol Endocrinol. 2009;7:40.

Rizzo P, Raffone E, Benedetto V. Effect of the treatment with myo-inositol plus folic acid plus melatonin in comparison with a treatment with myo-inositol plus folic acid on oocyte quality and pregnancy outcome in IVF cycles. A prospective, clinical trial. Eur Rev Med Pharmacol Sci. 2010;14:555–61.

Ronnberg L, Kauppila A, Leppaluoto J, Martikainen H, Vakkuri O. Circadian and seasonal variation in human preovulatory follicular fluid melatonin concentration. J Clin Endocrinol Metab. 1990;71:492–6.

Rossi T, Mazzilli F, Delfino M, Dondero F. Improved human sperm recovery using superoxide dismutase and catalase supplementation in semen cryopreservation procedure. Cell Tissue Bank. 2001;2:9–13.

Safarinejad MR. The effect of coenzyme Q(1)(0) supplementation on partner pregnancy rate in infertile men with idiopathic oligoasthenoteratozoospermia: an open-label prospective study. Int Urol Nephrol. 2012;44:689–700.

Saleh RA, Agarwal A, Nada EA, El-Tonsy MH, Sharma RK, Meyer A, Nelson DR, Thomas AJ. Negative effects of increased sperm DNA damage in relation to seminal oxidative stress in men with idiopathic and male factor infertility. Fertil Steril. 2003;79 Suppl 3:1597–605.

Seino T, Saito H, Kaneko T, Takahashi T, Kawachiya S, Kurachi H. Eight-hydroxy-2'-deoxyguanosine in granulosa cells is correlated with the quality of oocytes and embryos in an in vitro fertilization-embryo transfer program. Fertil Steril. 2002;77:1184–90.

Shahar S, Wiser A, Ickowicz D, Lubart R, Shulman A, Breitbart H. Light-mediated activation reveals a key role for protein kinase A and sarcoma protein kinase in the development of sperm hyper-activated motility. Hum Reprod. 2011;26:2274–82.

Shannon P. Factors affecting semen preservation and conception rates in cattle. J Reprod Fertil. 1978;54:519–27.

Sharma RK, Agarwal A. Role of reactive oxygen species in male infertility. Urology. 1996;48:835–50.

Sharma RK, Vemulapalli S, Kohn S, Agarwal A. Effect of centrifuge speed, refrigeration medium, and sperm washing medium on cryopreserved sperm quality after thawing. Arch Androl. 1997;39:33–8.

Sharma RK, Said T, Agarwal A. Sperm DNA damage and its clinical relevance in assessing reproductive outcome. Asian J Androl. 2004;6:139–48.

Shekarriz M, Dewire DM, Thomas Jr AJ, Agarwal A. A method of human semen centrifugation to minimize the iatrogenic sperm injuries caused by reactive oxygen species. Eur Urol. 1995;28:31–5.

Shen S, Khabani A, Klein N, Battaglia D. Statistical analysis of factors affecting fertilization rates and clinical outcome associated with intracytoplasmic sperm injection. Fertil Steril. 2003;79:355–60.

Shiva M, Gautam AK, Verma Y, Shivgotra V, Doshi H, Kumar S. Association between sperm quality, oxidative stress, and seminal antioxidant activity. Clin Biochem. 2011;44:319–24.

Sikka SC. Role of oxidative stress and antioxidants in andrology and assisted reproductive technology. J Androl. 2004;25:5–18.

Singh AK, Chattopadhyay R, Chakravarty B, Chaudhury K. Markers of oxidative stress in follicular fluid of women with endometriosis and tubal infertility undergoing IVF. Reprod Toxicol. 2013;42:116–24.

Squirrell JM, Lane M, Bavister BD. Altering intracellular pH disrupts development and cellular organization in preimplantation hamster embryos. Biol Reprod. 2001;64:1845–54.

Szymanski W, Kazdepka-Zieminska A. [Effect of homocysteine concentration in follicular fluid on a degree of oocyte maturity]. Ginekol Pol. 2003;74:1392–6.

Takenaka M, Horiuchi T, Yanagimachi R. Effects of light on development of mammalian zygotes. Proc Natl Acad Sci U S A. 2007;104:14289–93.

Tamura H, Takasaki A, Miwa I, Taniguchi K, Maekawa R, Asada H, Taketani T, Matsuoka A, Yamagata Y, Shimamura K, Morioka H, Ishikawa H, Reiter RJ, Sugino N. Oxidative stress impairs oocyte quality and melatonin protects oocytes from free radical damage and improves fertilization rate. J Pineal Res. 2008;44:280–7.

Tavilani H, Goodarzi MT, Vaisi-Raygani A, Salimi S, Hassanzadeh T. Activity of antioxidant enzymes in seminal plasma and their relationship with lipid peroxidation of spermatozoa. Int Braz J Urol. 2008;34:485–91.

Taylor K, Roberts P, Sanders K, Burton P. Effect of antioxidant supplementation of cryopreservation medium on post-thaw integrity of human spermatozoa. Reprod Biomed Online. 2009;18:184–9.

Thomson LK, Fleming SD, Aitken RJ, De Iuliis GN, Zieschang JA, Clark AM. Cryopreservation-induced human sperm DNA damage is predominantly mediated by oxidative stress rather than apoptosis. Hum Reprod. 2009;24:2061–70.

Tremellen K, Miari G, Froiland D, Thompson J. A randomised control trial examining the effect of an antioxidant (Menevit) on pregnancy outcome during IVF-ICSI treatment. Aust N Z J Obstet Gynaecol. 2007;47:216–21.

Tunc O, Thompson J, Tremellen K. Improvement in sperm DNA quality using an oral antioxidant therapy. Reprod Biomed Online. 2009;18:761–8.

Turi A, Giannubilo SR, Bruge F, Principi F, Battistoni S, Santoni F, Tranquilli AL, Littarru G, Tiano L. Coenzyme Q10 content in follicular fluid and its relationship with oocyte fertilization and embryo grading. Arch Gynecol Obstet. 2012;285:1173–6.

Unfer V, Raffone E, Rizzo P, Buffo S. Effect of a supplementation with myo-inositol plus melatonin on oocyte quality in women who failed to conceive in previous in vitro fertilization cycles for poor oocyte quality: a prospective, longitudinal, cohort study. Gynecol Endocrinol. 2011; 27:857–61.

Van Blerkom J, Davis PW. Cytogenetic, cellular, and developmental consequences of cryopreservation of immature and mature mouse and human oocytes. Microsc Res Tech. 1994;27:165–93.

Van Dyk Q, Lanzendorf S, Kolm P, Hodgen GD, Mahony MC. Incidence of aneuploid spermatozoa from subfertile men: selected with motility versus hemizona-bound. Hum Reprod. 2000;15:1529–36.

Vicari E, Calogero AE. Effects of treatment with carnitines in infertile patients with prostato-vesiculo-epididymitis. Hum Reprod. 2001;16:2338–42.

Wallock LM, Tamura T, Mayr CA, Johnston KE, Ames BN, Jacob RA. Low seminal plasma folate concentrations are associated with low sperm density and count in male smokers and nonsmokers. Fertil Steril. 2001;75:252–9.

Wang X, Sharma RK, Gupta A, George V, Thomas AJ, Falcone T, Agarwal A. Alterations in mitochondria membrane potential and oxidative stress in infertile men: a prospective observational study. Fertil Steril. 2003;80 Suppl 2:844–50.

Wang YX, Yang SW, Qu CB, Huo HX, Li W, Li JD, Chang XL, Cai GZ. [L-carnitine: safe and effective for asthenozoospermia]. Zhonghua Nan Ke Xue. 2010;16:420–2.

Will MA, Clark NA, Swain JE. Biological pH buffers in IVF: help or hindrance to success. J Assist Reprod Genet. 2011;28:711–24.

Wirleitner B, Vanderzwalmen P, Stecher A, Spitzer D, Schuff M, Schwerda D, Bach M, Schechinger B, Herbert Zech N. Dietary supplementation of antioxidants improves semen quality of IVF patients in terms of motility, sperm count, and nuclear vacuolization. Int J Vitam Nutr Res. 2012;82:391–8.

Wong WY, Merkus HM, Thomas CM, Menkveld R, Zielhuis GA, Steegers-Theunissen RP. Effects of folic acid and zinc sulfate on male factor subfertility: a double-blind, randomized, placebo-controlled trial. Fertil Steril. 2002;77:491–8.

Woods EJ, Benson JD, Agca Y, Critser JK. Fundamental cryobiology of reproductive cells and tissues. Cryobiology. 2004;48:146–56.

Zelen I, Mitrovic M, Jurisic-Skevin A, Arsenijevic S. Activity of superoxide dismutase and catalase and content of malondialdehyde in seminal plasma of infertile patients. Med Pregl. 2010;63:624–9.

Zini A, San Gabriel M, Baazeem A. Antioxidants and sperm DNA damage: a clinical perspective. J Assist Reprod Genet. 2009;26:427–32.

Zribi N, Feki Chakroun N, El Euch H, Gargouri J, Bahloul A, Ammar Keskes L. Effects of cryopreservation on human sperm deoxyribonucleic acid integrity. Fertil Steril. 2010;93:159–66.